BrightRED Revision

Advanced Higher CHEMISTRY

Archie Gibb and David Hawley

First published in 2010 by:

Bright Red Publishing Ltd
6 Stafford Street
Edinburgh
EH3 7AU

Reprinted (with corrections) in 2011

A CIP record for this book is available from the British Library

ISBN 978-1-906736-19-4

With thanks to Ken Vail Graphic Design, Cambridge (layout), Anna Clark (editorial)
Cover design by Caleb Rutherford – eidetic
Illustrations by Beehive Illustration (Mark Turner) and Ken Vail Graphic Design

Acknowledgements
Every effort has been made to seek all copyright-holders. If any have been overlooked, then Bright Red Publishing will be delighted to make the necessary arrangements. All Internet links in the text were correct at the time of going to press.

Permission has been sought from all relevant copyright holders and Bright Red Publishing is grateful for the use of the following:

A photograph © UHB Trust/Getty Images (p 83);
A graph taken from the 2000 Advanced Higher Chemistry CSYS Exam Paper © Scottish Qualifications Authority (p 13).

A table taken from page 54 of 'Electronic Structure and the Periodic Table' (http://www.ltscotland.org.uk/nationalqualifications/resources/c/nqresource_tcm4228670.asp) published by Learning and Teaching Scotland, August 2000, now published by Education Scotland © Crown Copyright

Printed and bound in the UK by WG Baird Limited.

CONTENTS

ADVANCED HIGHER CHEMISTRY

COURSE STRUCTURE

The Advanced Higher Chemistry course is divided into four units:

- Unit 1: Electronic Structure and the Periodic Table (½ unit)
- Unit 2: Principles of Chemical Reactions
- Unit 3: Organic Chemistry
- Unit 4: Chemistry Investigation (½ unit)

ASSESSMENT

There are two types of assessment – **internal** and **external**.

INTERNAL ASSESSMENT

For each of the first three units, the **internal assessment** consists of a National Assessment Bank (NAB) test. You have to gain at least half marks in order to pass. Practical abilities are also assessed internally; you have to write a report to a satisfactory standard on any **one** of the 12 Prescribed Practical Activities (PPAs).

The internal assessment of the Investigation requires you to keep a Day Book. This is an on-going record of your planning, as well as the place that you collect and analyse your experimental data.

EXTERNAL ASSESSMENT

The **external assessment** consists of two parts – the **written examination** and the **Investigation Report**.

The written examination consists of one paper of duration 2 hours and 30 minutes, with a total allocation of **100 marks**. The examination paper is divided into two sections:

- **Section A**, worth **40 marks**, is made up of 40 multiple-choice questions.
- **Section B**, worth **60 marks**, contains questions that require written answers. In this section of the paper, approximately 6 marks are allocated to questions based on any of the 12 PPAs of the course.

Of the 100 marks in the paper, between 50 and 55 marks are allocated to Knowledge and Understanding (KU) questions and between 45 and 50 marks to Problem-Solving (PS) questions. Also, **25 marks** are allocated to the Investigation Report, giving a total of 125 marks for the external assessment.

The course award is graded A, B, C or D depending on how well you do in the written examination **and** the Investigation Report, i.e. your total mark out of 125. In order to gain the course award, you must also pass all aspects of the internal assessment.

STRUCTURE AND AIM OF THIS BOOK

The aim of this revision guide is to help you achieve success in the final examination by providing you with a concise and engaging coverage of the subject arrangements.

This book addresses the first three units of the course and within each unit there is a double-page spread on each of the sub-sections. In the Arrangements document, there is a sub-section in Unit 3 labelled 'Permeating aspects of organic chemistry' but in this book, there is no separate spread devoted exclusively to this. As the label makes clear, the content of this sub-section permeates Unit 3 and is addressed at appropriate points within the Unit 3 spreads.

Each double-page spread:

- covers the content of the sub-section in a logical and digestible manner and will allow you to gain a good understanding of the key ideas and concepts.
- contains **'Don't Forget'** sections. These flag up vital pieces of knowledge that you need to remember and important things that you must be able to do.
- includes a **'Let's think about this'** section. Some have questions that are designed to test your knowledge and understanding of the content. Others are designed to extend your knowledge of the subject and provide additional interest. Answers are also provided, either immediately after the questions or on page 99.

At the end of each unit there is a section devoted to the PPAs of that unit. You are taken through the 'Aim', 'Procedure', 'Results' and 'Conclusion'. There is also an 'Evaluation' in which important aspects of the PPAs are highlighted.

The book ends with a section offering advice on the production of the Investigation Report (see pages 94 to 98).

ELECTROMAGNETIC SPECTRUM AND ASSOCIATED CALCULATIONS

ELECTROMAGNETIC RADIATION

Since chemical reactions involve the reacting atoms, molecules or ions colliding together it is their outer electrons that are involved when different substances react together. Therefore, it is necessary to understand the electronic structure of an atom of an element to explain the chemical properties of that element. Much of the information about electronic structure comes from spectroscopy and, for an understanding of this, it is necessary to consider electromagnetic radiation.

Radiation such as light, microwaves, X-rays and television signals is known as electromagnetic radiation. Electromagnetic radiation can be considered in terms of waves which travel in a vacuum at a constant speed of $3{\cdot}00 \times 10^8\,\mathrm{m\,s^{-1}}$ and which have wavelengths of between approximately 10^{-14} and 10^4 metres.

Different types of electromagnetic radiation make up the electromagnetic spectrum. Visible light – the radiation that our eyes can detect – makes up only a small part of the electromagnetic spectrum.

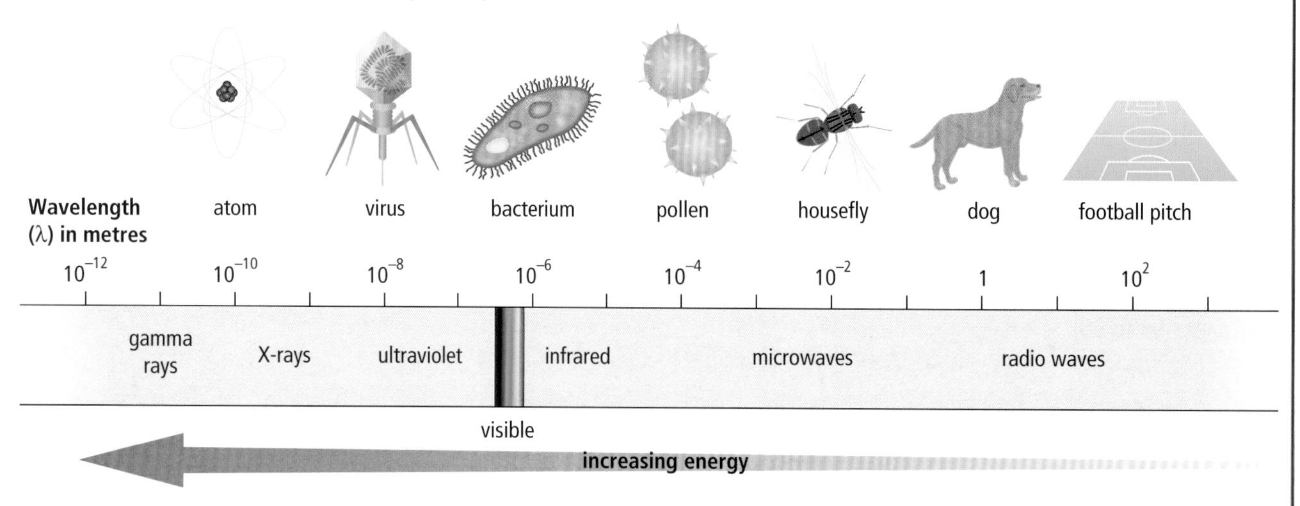

The diagram shows some of the different types of radiation that make up the electromagnetic spectrum. At the high-energy end of the spectrum, the waves are so tightly packed that they are closer together than the size of an atom, whereas at the low-energy end the waves are further apart than the length of a football pitch.

WAVELENGTH, FREQUENCY AND VELOCITY OF ELECTROMAGNETIC RADIATION

Electromagnetic radiation can be specified by its wavelength and by its frequency.

- **Wavelength** is the distance between adjacent crests or high points of a wave. This distance can be measured in metres. However, due to the types of radiation considered in chemistry, very often the unit of wavelength is the nanometre (nm). One nanometre is 10^{-9} metres. The symbol for wavelength is the Greek letter lambda, λ.
- **Frequency** is the number of wavelengths that pass a fixed point in one second. The symbol for frequency is the Greek letter nu, ν. Frequency is measured as $\frac{1}{\text{time}}$ and so has the unit $\mathrm{s^{-1}}$. This unit is now more commonly known as the Hertz, Hz.

All electromagnetic radiation travels at the same **velocity** in a vacuum. This is equal to $3{\cdot}00 \times 10^8\,\mathrm{m\,s^{-1}}$. This value is given the symbol, c, and in Advanced Higher Chemistry we take it to be the constant speed of electromagnetic radiation anywhere.

contd

WAVELENGTH, FREQUENCY AND VELOCITY OF ELECTROMAGNETIC RADIATION contd

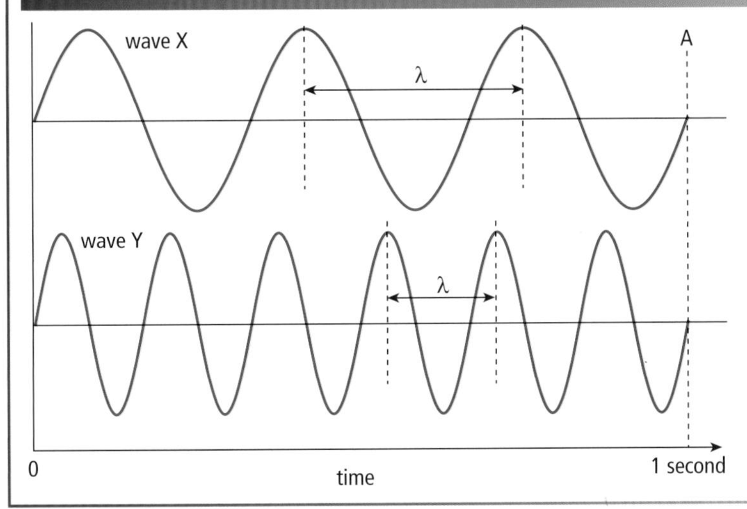

The wavelength of wave X has double the value of the wavelength of wave Y. Since both waves will be travelling at the same velocity ($c = 3 \times 10^8\,m\,s^{-1}$), then twice as many wavelengths of wave Y will pass position A every second compared to wave X. This means that the frequency of wave Y is twice that of wave X.

DON'T FORGET

The relationship between velocity, wavelength and frequency is $c = \lambda\nu$. You must know this relationship and be able to use it in calculations.

ENERGY ASSOCIATED WITH ELECTROMAGNETIC RADIATION

- Electromagnetic radiation with **short wavelength and high frequency**, such as gamma rays, is at the **high-energy end of the electromagnetic spectrum**.
- Radio waves and other radiation with **long wavelengths and low frequencies** are at the **low-energy end of the electromagnetic spectrum**.

Electromagnetic radiation is considered to have a **dual nature**, based on wave motion *and* movement of particles. It can be considered as a stream of particles called photons, but with wave properties such as frequency and wavelength.

The energy, E, carried by a photon is related to its frequency by the equation, $E = h\nu$ in which h is known as Planck's constant and has the value $6{\cdot}63 \times 10^{-34}\,J\,s$.

Planck's constant is given on page 19 of the SQA Data Booklet which can be downloaded from http://www.sqa.org.uk/files_ccc/NQChemistryDataBooklet_H_AdvH.pdf

In chemistry, energy values are normally calculated in $kJ\,mol^{-1}$ and so the above equation is more usefully written as $E = \frac{Lh\nu}{1000}$ where L = Avogadro constant ($6{\cdot}02 \times 10^{23}\,mol^{-1}$) and dividing by 1000 converts the value from J into kJ. The equation now relates the energy of one mole of photons to their frequency and the energy value is calculated in $kJ\,mol^{-1}$.

It is often useful to relate energy, in $kJ\,mol^{-1}$, to wavelength and the equation which can be used to do this is $E = \frac{Lhc}{1000\lambda}$

DON'T FORGET

You must know and be able to use these relationships. Exam papers often ask students to calculate the energy, in $kJ\,mol^{-1}$, associated with radiation of a particular wavelength given in nm.

LET'S THINK ABOUT THIS

Velocity is measured in $m\,s^{-1}$, wavelength is measured in m and frequency in s^{-1}.

So, when units are substituted into the equation, $c = \lambda\nu$, the value for c is given in $m\,s^{-1}$ and the value for $\lambda\nu$ is also in $m\,s^{-1}$, which makes sense. However in most calculations, the wavelength is given in nm (10^{-9} m). So, you must remember to convert the wavelength into metres when you are using this equation.

For example, if you have to calculate the frequency of light which has wavelength 509 nm, you would use the value 509×10^{-9} m as the wavelength. Try this calculation; you should get a frequency of $5{\cdot}89 \times 10^{14}\,s^{-1}$ or $5{\cdot}89 \times 10^{14}$ Hz.

Now try to calculate the energy, in $kJ\,mol^{-1}$, corresponding to this wavelength or frequency.

ELECTRONIC CONFIGURATION AND THE PERIODIC TABLE

EMISSION SPECTRA

A continuous spectrum in the visible region of the electromagnetic spectrum is observed when white light (such as light from an ordinary light bulb) is passed through a glass prism or a diffraction grating. A continuous spectrum is like a rainbow.

When a high voltage is passed through a tube of hydrogen gas at low pressure, a coloured light is produced. When this light is analysed by passing it through a diffraction grating or a spectroscope, only coloured lines corresponding to certain frequencies or wavelengths are seen.

Different patterns of coloured lines are seen when a high voltage is passed through other gases at low pressure and the coloured light is analysed; different gases produce lines at different frequencies and wavelengths. These are known as emission spectra and every element produces a unique emission spectrum, i.e. the wavelengths of the lines that are produced are unique to that element and different to the lines produced by other elements.

The lines seen in emission spectra are due to photons of energy that are emitted (given out) when electrons in higher energy levels move to lower energy levels. Lines corresponding to definite energy values are observed and this provides evidence that the transfer of energy is in small fixed amounts which are known as **quanta**. The idea that photons 'carry' definite quantities of energy forms the basis of quantum theory.

EMISSION SPECTRUM OF HYDROGEN

Hydrogen is the simplest element, having only one electron. This suggests that its emission spectrum should be the easiest to interpret. Although lines are produced in the infra-red and ultraviolet parts of the electromagnetic spectrum, we will first consider the emission spectrum of hydrogen in the visible part of the electromagnetic spectrum.

When an electron **absorbs** the appropriate quantity of energy it moves from the ground state to a higher energy level. We say that the electron is now in an **excited** state. Each line in the **emission** spectrum corresponds to the energy **given out** by an excited electron when it moves to a state of lower energy. This can be either the ground state or a lower excited state, as shown in the diagram on the next page.

The lines seen in the visible region in the emission spectrum of hydrogen are due to excited electrons 'falling back' to the n = 2 energy level, i.e. to the second shell. This is known as the Balmer series after the person who first discovered and explained these lines.

Although hydrogen is the simplest element the emission spectrum is still very complicated. There are other series of lines corresponding to excited electrons falling back to the n = 5, n = 4, n = 3 and n = 1 energy levels. These different series are shown in the table opposite and three of them are shown in the energy shell diagram which is not drawn to scale. The series are named after their discoverers.

contd

EMISSION SPECTRUM OF HYDROGEN contd

The two series that we will concentrate on are the Balmer series, since it is in visible region, and the Lyman series which involves electrons falling back to the n = 1 ground state. Since higher quantities of energy are released when electrons move back to the n = 1 energy level compared to other energy levels, the Lyman series appears in the ultraviolet part of the electromagnetic spectrum. The Lyman series is not visible with the naked eye.

Name of series	Energy level to which the excited electron falls	Where the lines are produced
Lyman	n = 1	ultraviolet region
Balmer	n = 2	visible region
Paschen	n = 3	infra-red
Brackett	n = 4	infra-red
Pfund	n = 5	infra-red

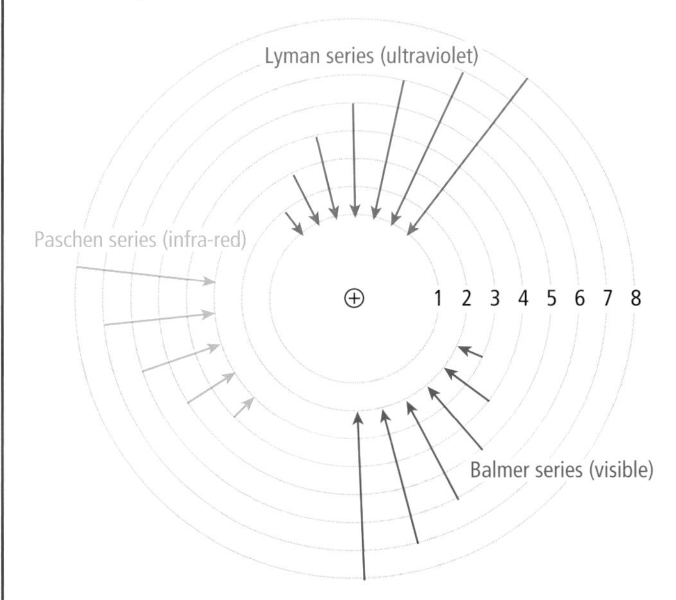

These electronic transitions, shown as coloured arrows, correspond to lines with definite values of frequency and wavelength. The diagram is not to scale and the blue lines going back to the ground state should be much longer than the other lines because the difference between n = 1 and n = 2 energy levels is by far the greatest, followed by the difference between n = 2 and n = 3 etc.

The diagram shows the electronic transitions that give rise to the Lyman, Balmer and Paschen series. Each line in the emission spectra is due to a transition between definite energy levels. Using the frequency or wavelength of a line, it is possible to calculate the energy difference between energy levels.

The structure of the atom as drawn above with electrons orbiting a positive nucleus was proposed by the Danish Scientist, Neils Bohr. It worked well for hydrogen but is not a good model for atoms of other elements.

DON'T FORGET

The lines in an emission spectrum give information about the difference in energy levels when an excited electron moves to a lower energy level and energy is released.

LET'S THINK ABOUT THIS

The energy levels get closer and closer together the further they are from the nucleus. If you consider the emission spectrum of hydrogen at the high-energy or short-wavelength end, the lines get closer and closer together until we say that they converge. This is known as the convergence limit. The convergence limit in the Lyman series for hydrogen is at 91·2 nm.

You can think of this as the excited electron being at its highest energy level and losing energy as it drops down to the ground state. The highest energy level is as far as the electron can be from the nucleus without being completely removed from the hydrogen atom. Consider this happening in reverse, so that the electron moves from the ground state to just beyond the highest energy level – ionisation would have occurred.

Now consider 1 mole of atoms in the gas state forming 1 mole of ions in the gas state. The energy required to do this would be the ionisation energy of hydrogen. If you calculate the energy, in kJ mol^{-1}, corresponding to the wavelength of the convergence limit in the Lyman series, you should find that it is almost exactly the same as the value for the ionisation energy of hydrogen given in the SQA Data Booklet. Try this for yourself.

You can get more information on the emission spectra of hydrogen at http://csep10.phys.utk.edu/astr162/lect/light/absorption.html

HEISENBERG, PAULI AND SHAPES OF ORBITALS

SIMPLE QUANTUM MECHANICS

Bohr's simple model of an atom only worked with hydrogen atoms. A new science known as quantum mechanics was required to explain atoms with more than one electron. Quantum mechanics is beyond the scope of this book but, in simple terms, it considers electrons both as particles and as waves. When the mathematics involved with quantum mechanics is used, the results are fairly easy to interpret. Electrons within atoms are said to be quantised. This means that they can only possess fixed amounts of energy known as quanta. As a result, electrons can be defined in terms of quantum numbers.

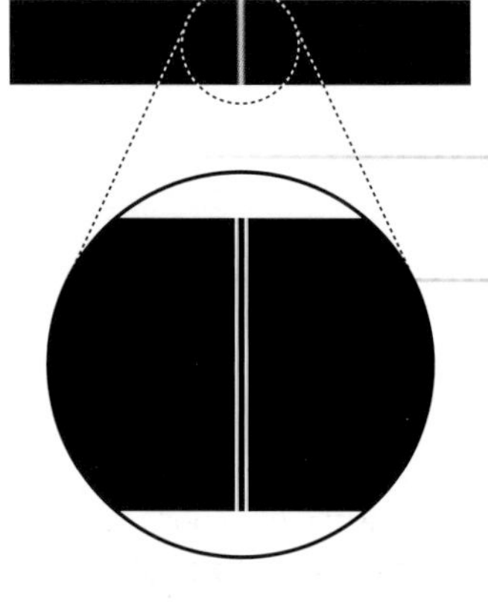

The high resolution emission spectrum of sodium suggests that shells are subdivided into ***subshells****. Using quantum mechanics, it can be calculated that all shells have at least one subshell.*

SHELLS, SUBSHELLS AND QUANTUM NUMBERS

As you already know, electrons are arranged in shells. Each shell has a number. For the first shell n = 1, for the second shell n = 2, and so on. This number we now call the **principal quantum number, n**. The first shell is the shell closest to the nucleus. The higher the value of n, the further the electrons within that shell are from the nucleus.

If emission spectra of elements other than hydrogen are studied using high resolution spectroscopes, the single lines seen at low resolution often become two or three lines very close together. For example, in the emission spectrum of sodium at low resolution one yellow line is seen. However, under higher resolution it can be shown that, in fact, there are two lines very close together. This suggests that shells are divided into subshells.

Subshells are labelled s, p, d and f. These letters originally came from old spectroscopic terms 'sharp', 'principal', 'diffuse' and 'fundamental' but these terms are no longer important.

With the exception of hydrogen, the subshells within each shell have slightly different energies; the s-subshell has the lowest energy, then p, then d and so on. The table below shows the different subshells present in each shell. Each type of subshell contains one or more **orbitals**.

Shell	Subshells
first	1s
second	2s, 2p
third	3s, 3p, 3d
fourth	4s, 4p, 4d, 4f

In the past, scientists considered electrons as particles with a negative charge and with almost zero mass. Like photons, however, electrons can also be thought of as having the properties of both waves and particles. For example, in an electron microscope, a beam of electrons of particular wavelength is fired at the specimen to produce an image. This suggests that the beam of electrons is behaving like a beam of light waves.

Heisenberg's uncertainty principle

Heisenberg's uncertainty principle states that 'it is impossible to define with absolute precision, simultaneously, both the position and momentum (or velocity) of an electron'. So, electrons can be described in terms of probability rather than by definite position. In turn, an atomic orbital is defined as a region in space in which the probability of finding an electron is high, say over 90%.

As we have seen, the principal quantum number, n, describes the shell occupied by the electron. There are other quantum numbers which give some finer details but are not so important at Advanced Higher level. Here is some information about them.

- The angular momentum quantum number (l) gives the shape of the orbital.
- The magnetic quantum number (m_l) provides information about the number of orbitals and their orientation in space.
- The spin quantum number (m_s) determines the direction of spin and has the value $+\frac{1}{2}$ or $-\frac{1}{2}$, meaning that electrons can be considered as spinning in either of two directions.

contd

SHELLS, SUBSHELLS AND QUANTUM NUMBERS contd

The Pauli exclusion principle

The Pauli exclusion principle states that no two electrons in an atom can have the same four quantum numbers. Put simply, this means that an atomic orbital can hold a maximum of two electrons and these two electrons must be spinning in opposite directions.

NUMBERS AND SHAPES OF ORBITALS

An **s orbital** is spherical like a ball. The s orbital in the first shell is smaller than the s orbital in the second shell. The s orbital in the third shell is bigger yet, and so on.

1s 2s 3s

Since s orbitals are spherical, they are non-directional.

Since the first shell has only two electrons it has only one orbital – the 1s orbital. The second and subsequent shells all have **p orbitals**, as well as an s orbital. The p subshell has three different p orbitals of the same energy. Orbitals which have the same energy are said to be **degenerate**.

You must know the shapes of the s and p orbitals and be able to draw them. You must also remember that the maximum number of electrons in any orbital is two.

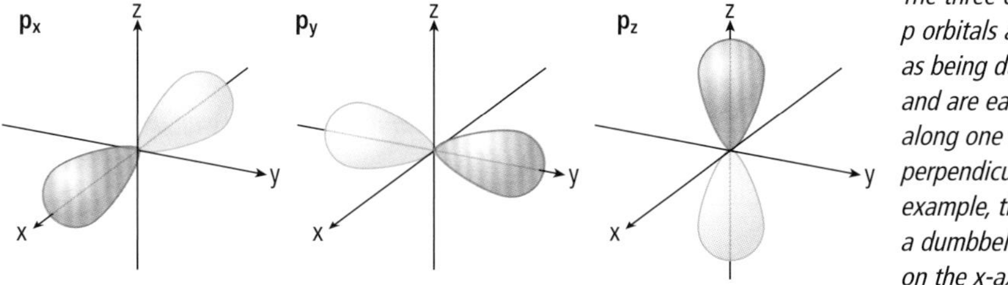

The three degenerate p orbitals are described as being dumbbell shaped and are each aligned along one of the three perpendicular axes. For example, the p_x orbital has a dumbbell shape and sits on the x-axis.

The second shell can hold a maximum of eight electrons. The s orbital can hold two of these electrons and the three different p orbitals can hold two electrons each. Therefore the second shell has only one s and three p orbitals.

The third shell can hold a maximum of 18 electrons. Two of these electrons are in the s orbital and the three p orbitals hold another six electrons. The remaining 10 electrons are accommodated in d orbitals. Since an orbital can hold a maximum of two electrons, there must be five d orbitals in the third and subsequent shells.

You must be able to recognise the d orbitals from diagrams and realise that the d_{z^2} has a different shape from the others.

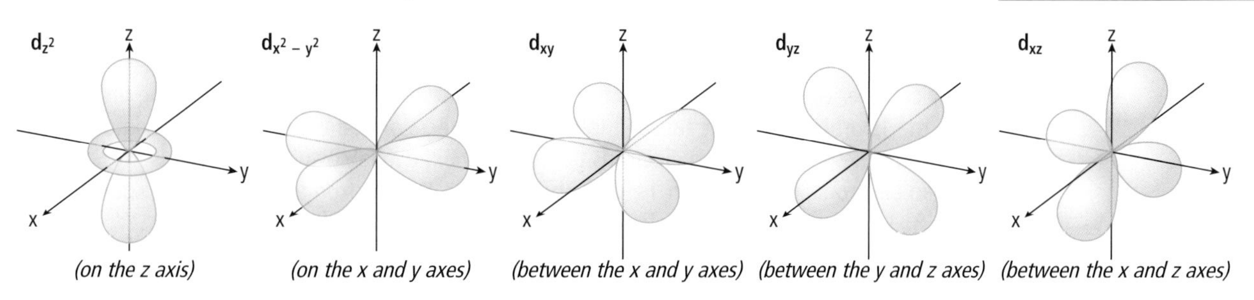

(on the z axis) (on the x and y axes) (between the x and y axes) (between the y and z axes) (between the x and z axes)

These five d orbitals are degenerate with each other but have higher energies than the s and p orbitals in the same shell. Four of the d orbitals are double dumbbell in shape.

LET'S THINK ABOUT THIS

The maximum number of electrons in the fourth and subsequent shells is 32.

How many electrons will be left after the s, p and d orbitals are filled? These electrons will occupy the f subshell.

You should be able to calculate the number of f orbitals in the f subshell. Remember that each orbital can hold no more than two electrons. By the way, the f orbitals are even more complicated than the d orbitals, but you do not have to know or recognise their shapes.

WRITING ELECTRONIC CONFIGURATIONS

THE AUFBAU PRINCIPLE AND HUND'S RULE

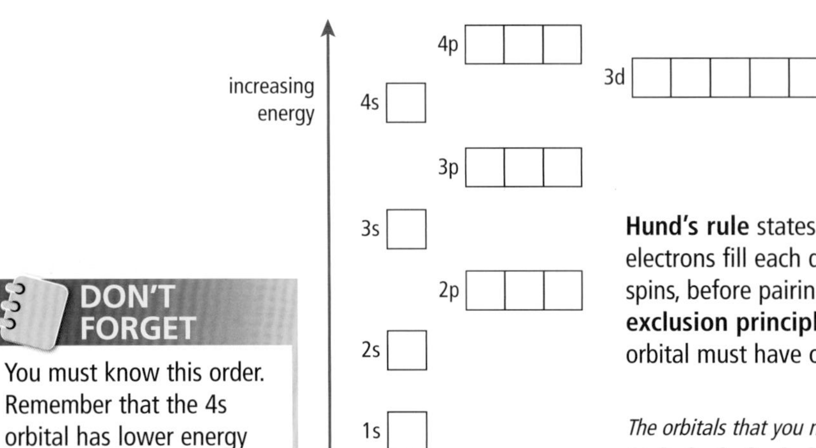

Electrons fill the orbitals in order of increasing energy, meaning that the lowest energy subshells are filled first. This is known as the **aufbau principle**. Of course, some subshells such as the p subshell and the d subshell have degenerate orbitals.

Hund's rule states that when degenerate orbitals are available, electrons fill each degenerate orbital singly and with parallel spins, before pairing up to fill the orbitals. To fit in with the **Pauli exclusion principle** (see page 9), two electrons in the same orbital must have opposite spins.

The orbitals that you need to know, in order of increasing energy, are 1s, 2s, 2p, 3s, 3p, 4s, 3d, 4p. This is shown diagrammatically on the left with boxes representing orbitals.

DON'T FORGET

You must know this order. Remember that the 4s orbital has lower energy than the 3d orbitals.

ELECTRONIC CONFIGURATIONS USING SPECTROSCOPIC NOTATION

Electronic configurations are similar to electron arrangements. However, electronic configurations show the subshells that the electrons are in, whereas electron arrangements show only the number of electrons in each shell.

For example, lithium has an electron arrangement **2, 1** but its electronic configuration is $1s^2\ 2s^1$

The characters in red indicate the shell and subshell. The numbers in **blue** indicate the number of electrons in that subshell. So the two electrons which lithium has in the first shell are located in the 1s subshell or 1s orbital. The one electron in lithium's second shell is in the 2s subshell or 2s orbital.

Now consider carbon. It has the electron arrangement **2, 4**. The two electrons in the first shell go into the 1s orbital. The next subshell to be filled is the 2s orbital which holds a maximum of two electrons. The remaining two electrons go into the next available subshell which is 2p. So, carbon has an electronic configuration $\mathbf{1s^2\ 2s^2\ 2p^2}$. Likewise, the electron arrangement of sodium which is **2, 8, 1** can be written as an electronic configuration of $\mathbf{1s^2\ 2s^2\ 2p^6\ 3s^1}$.

ELECTRONIC CONFIGURATIONS USING ORBITAL BOX NOTATION

Now consider the electronic configuration of carbon again – it is $1s^2\ 2s^2\ 2p^2$. Remember, there are three different p orbitals in the 2p subshell: the p_x orbital lies on the x-axis; the p_y orbital lies on the y-axis; and the p_z orbital lies on the z-axis. The different p orbitals are degenerate. To obey Hund's rule these degenerate orbitals must be filled singly before spin pairing occurs and to obey the Pauli exclusion principle, when an orbital is full with two electrons, these electrons must have opposite spins. This is not shown using spectroscopic notation but is seen when orbital box notation is used.

Orbital box notation shows the orbitals as boxes with arrows representing electrons. Arrows pointing in the same direction show electrons with the same spin. Electrons with opposite spin are represented by arrows pointing in opposite directions.

contd

ELECTRONIC CONFIGURATIONS USING ORBITAL BOX NOTATION contd

The diagram below shows the electronic configuration for carbon in orbital box notation. The two electrons in the p subshell are in different orbitals but have parallel spins and the electrons sharing the same orbitals in the 1s and 2s subshells have opposite spins. The diagram also suggests that one of the 2p orbitals is empty. In reality, there is no such thing as an empty orbital. If an orbital is empty then it does not exist. However, it is acceptable to show 'empty orbitals' in this type of notation.

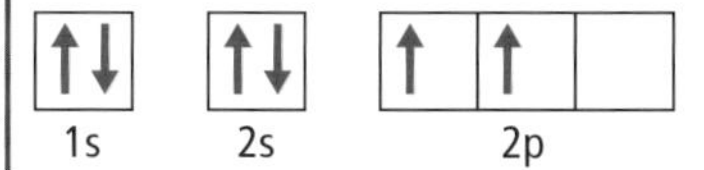

The electron arrangement of potassium is given in the SQA Data Booklet as 2, 8, 8, 1. In spectroscopic notation, the electronic configuration of potassium is $1s^2\ 2s^2\ 2p^6\ 3s^2\ 3p^6\ 4s^1$. The diagram below shows this information presented in orbital box notation.

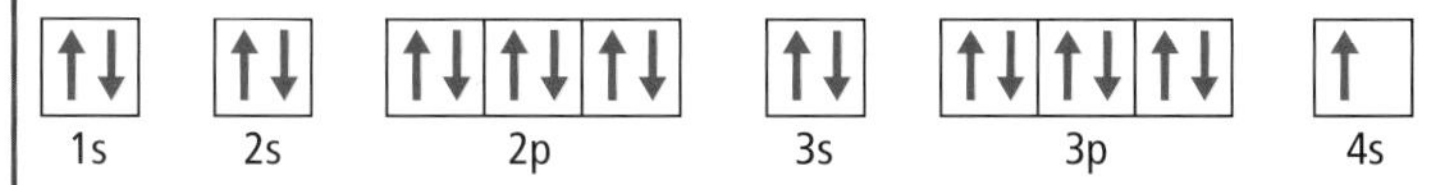

The electron arrangement of titanium is given in the SQA Data Booklet as 2, 8, 10, 2. In spectroscopic notation, the electronic configuration of titanium is $1s^2\ 2s^2\ 2p^6\ 3s^2\ 3p^6 3d^2\ 4s^2$. The diagram below shows this information presented in orbital box notation.

1s 2s 2p 3s 3p 3d 4s

DON'T FORGET

For every element, the electronic configuration must agree with the electron arrangement as given in the SQA Data Booklet. Looking at the electron arrangements in the Data Booklet, you can see that there should be two electrons in the 4s orbital before the 3d subshell starts to fill. You should be able to write the electronic configurations for all the elements up to krypton, atomic number 36.

LET'S THINK ABOUT THIS

1 Using both spectroscopic notation and orbital box notation, write down the electronic configurations of the following species – note that some are ions.

(a) He **(b)** N **(c)** Al^{3+} **(d)** Ar **(e)** Ca
(f) Ca^{2+} **(g)** Ni **(h)** Mn **(i)** Br^- **(j)** S^{2-}

2 Look at the graph of first ionisation energy plotted against atomic number for elements 1 to 36. You can see that first ionisation energy is a periodic property, since the pattern from Li to Ne is repeated from Na to Ar. From your work at Higher, can you recall the main factors that affect the ionisation energy of an element? In general, ionisation energy increases across a period (for example from Li to Ne) and decreases down a group, so from He to Kr or from Li to Rb.

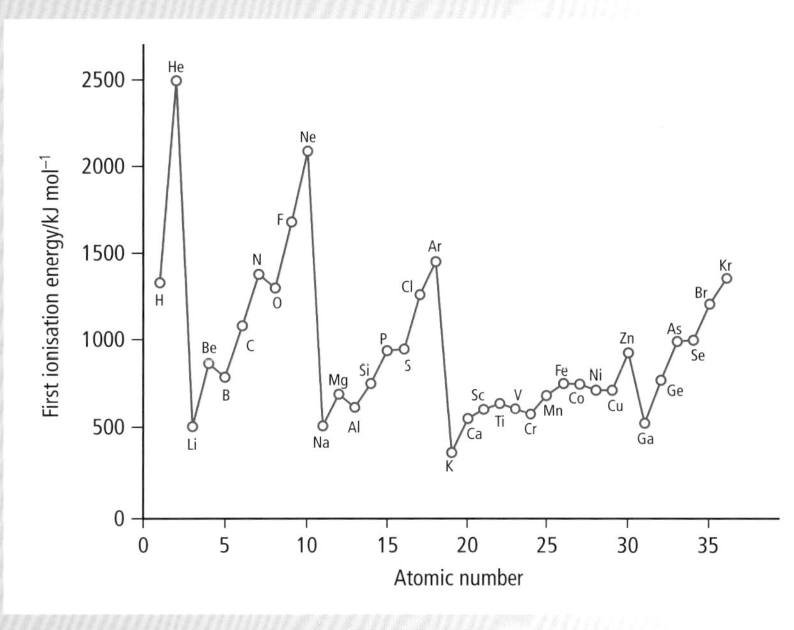

However, you can see that ionisation energies do not increase smoothly along a period; there is a drop from Be to B and from N to O. This is good evidence for subshells within each shell. Try to explain in terms of the relative stability of their electronic configurations why the first ionisation energy of B is less than that of Be, and why the first ionisation energy of O is less than that of N.

A helpful website is http://www.webchem.net/notes/Periodicity/ionisation_energy.htm

SPECTROSCOPY

ATOMIC EMISSION SPECTROSCOPY (AES)

Atomic emission spectroscopy is one of the oldest instrumental techniques for chemical analysis, and is still widely used for rapid identification and analysis of elements contained within a sample. It involves transitions between electronic energy levels in atoms or ions. These energy differences usually fit into the visible region of the electromagnetic spectrum (within the wavelength range 400–700 nm), but in some cases the energy difference is greater and falls within the ultraviolet region (wavelength range approximately 200–400 nm).

In AES the sample is first converted into a gas and then excited using a flame or electricity. The excited gas atoms emit energy in the form of light. Measuring the intensity of this light and analysing it for the presence of lines at definite wavelengths means that, not only can AES determine which elements are present in the sample, but can also do this quantitatively.

ATOMIC ABSORPTION SPECTROSCOPY (AAS)

Atomic absorption spectroscopy is also a technique for determining the concentration of a particular metal element in a sample. In AAS the electrons are promoted to higher energy levels by absorbing energy. Like AES, the energy differences are usually within the wavelength range 200–700 nm, meaning that ultraviolet and visible light is used. The wavelength of the absorbed light is specific to a particular element, so AAS is also used to determine which element is present. The intensity of absorbed light is used to determine the concentration of the element; AAS can be used to determine the concentration of over 60 different metals in solution.

USES OF AES AND AAS

Spectra can be used to give information about how much of a species is present in a sample. For example, the concentration of lead or aluminium in drinking water or in foodstuffs can be found, as can the quantity of different metals present in effluent water from industrial plants.

Using AAS, a calibration graph is first prepared from known concentrations of solutions of the metal in question. The radiation absorbed by these samples is plotted against concentration so when the unknown sample is analysed the concentration of that particular metal can be found from the graph. (This technique is very similar to that in Unit 1 PPA2, Colorimetric determination of manganese in steel – see page 29.)

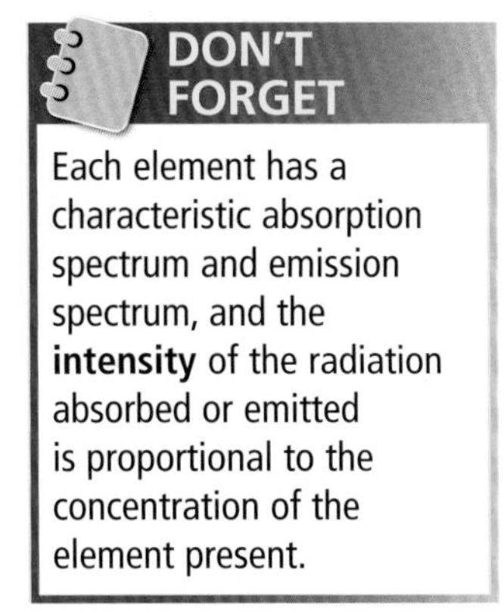

Each element has a characteristic absorption spectrum and emission spectrum, and the **intensity** of the radiation absorbed or emitted is proportional to the concentration of the element present.

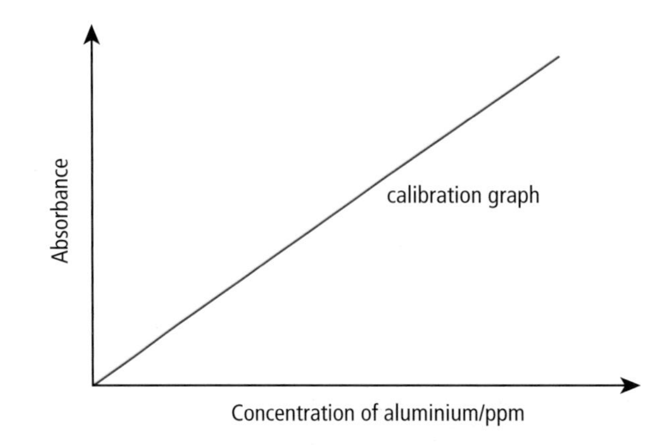

Absorbance values at the wavelength corresponding to aluminium are found using different solutions of known aluminium concentrations. A calibration graph is plotted. The percentage absorbance in a sample of drinking water is then found and the concentration of aluminium corresponding to this absorbance is read from the calibration graph.

contd

USES OF AES AND AAS contd

If you have a Scholar user name and password, you will find more information on atomic spectroscopy by logging on and going to Unit 1 Topic 1·4, Using spectra to identify samples.

Some webpages which give more information on AES and AAS include:
http://www.andor.com/learning/applications/Atomic_Spectroscopy/
http://elchem.kaist.ac.kr/vt/chem-ed/spec/atomic/aes.htm
http://elchem.kaist.ac.kr/vt/chem-ed/spec/atomic/aa.htm
You will also find a powerpoint presentation at http://webpage.pace.edu/dnabirahni/rahnidocs/Atomic%20Emission%20Spectroscopy.ppt

LET'S THINK ABOUT THIS

1 In 1904, Sir William Ramsay was the first Briton to be awarded the Nobel Prize for Chemistry. He was educated at Glasgow Academy and Glasgow University, and the prize was in recognition for the work he had done with others to discover the noble gases. Helium was discovered in the Sun before it was ever found on planet Earth. How do you think it was possible to do this?

You will find more information about Sir William Ramsay and his discoveries at http://nobelprize.org/nobel_prizes/chemistry/laureates/1904/ramsay-bio.html and http://www.answers.com/topic/william-ramsay

2 The concentration of calcium ions in solution can be determined using an atomic absorption spectrophotometer operating at wavelength 422·7 nm. The absorbance was measured for a range of known concentrations of calcium ions and the calibration graph shown below was drawn.

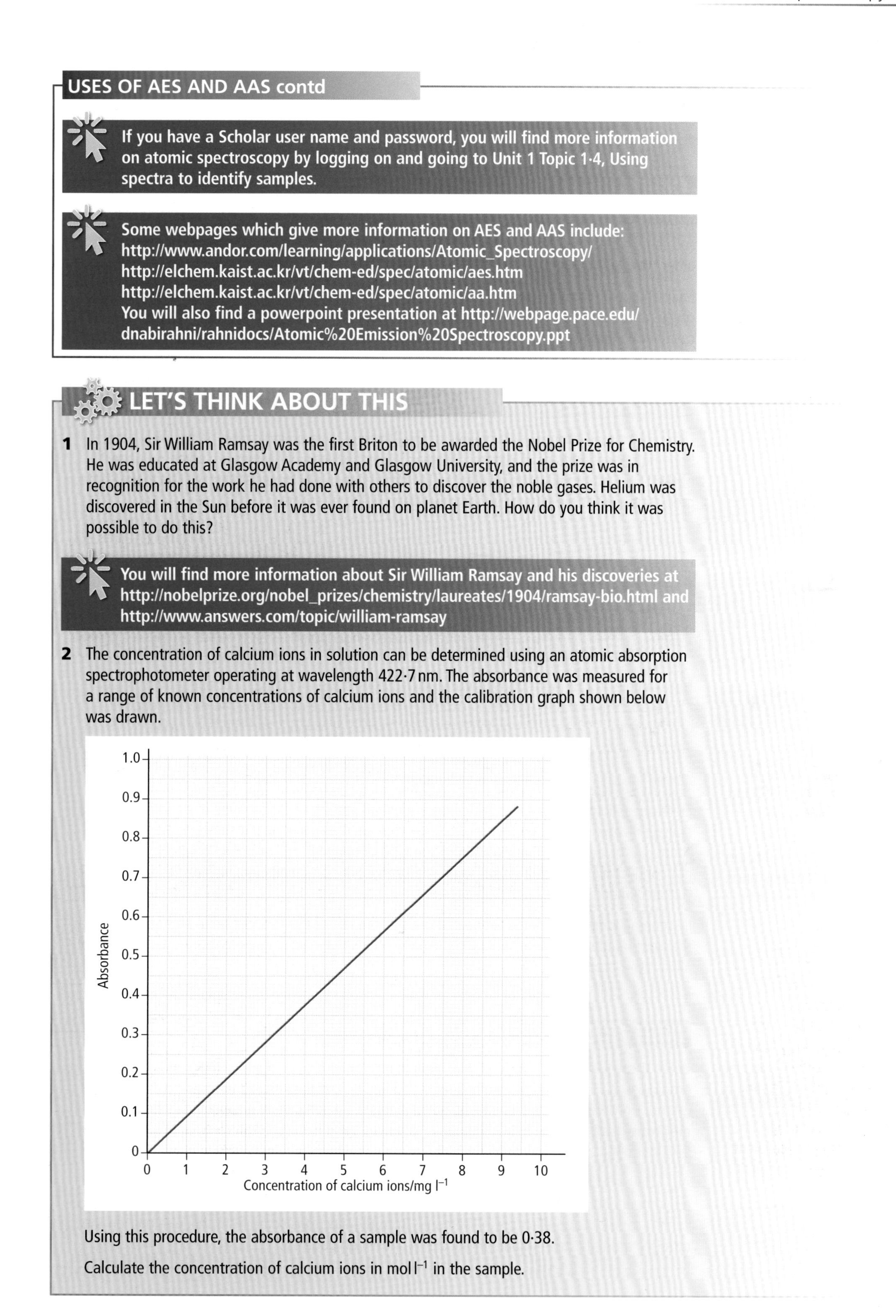

Using this procedure, the absorbance of a sample was found to be 0·38.

Calculate the concentration of calcium ions in mol l^{-1} in the sample.

CHEMICAL BONDING

TYPES OF BONDING

Types of bonding include metallic, non-polar covalent (or pure covalent bonding), polar covalent and ionic bonding. In this section we will not cover metallic bonding, but will concentrate on the other types of bonding.

Ionic and polar covalent bonding do not exist in elements. We can consider non-polar (pure) covalent bonding and ionic bonding as being at opposite ends of a bonding continuum. Polar covalent bonding sits between these two extremes.

ELECTRONEGATIVITY AND BONDING

Differences in electronegativity values of the elements give an indication of the likely type of bonding between atoms of different elements. Electronegativity values are given on page 10 of the SQA Data Booklet. Using ΔEN as the difference in electronegativity, we can make some predictions about the type of bonding between different atoms.

$\Delta EN = 0$		max $\Delta EN = 3{\cdot}2$
non-polar covalent	polar covalent	ionic

———————— ΔEN increasing ————————→

———————— bonding increasingly polar ————————→

Electronegativity differences are useful in predicting the type of bond formed, but you have a fuller picture when you have information about the properties of the substance.

For example, in PH_3, both phosphorus and hydrogen have the same electronegativity value of 2·2. So, $\Delta EN = 0$ and we would expect P–H bonds to be non-polar. The electronegativity of hydrogen is 2·2 and that of chlorine is 3·0. In HCl, $\Delta EN = 0{\cdot}8$ and the bonding in HCl is polar covalent.

As you are aware, the bonding in sodium chloride is ionic. ΔEN in NaCl is 2·1.

Electronegativity values and ΔEN values are useful predictors for the type of bonding but this doesn't always give the correct result. Properties such as whether the substance is an electrical conductor give definite evidence about the type of bonding. For example, covalent substances do not conduct when solid or liquid. Ionic substances are non-conductors in the solid state but do conduct when liquid.

In general, ionic bonds form between metals at the left of the Periodic Table and non-metals at the right-hand side. Covalent bonding usually occurs between non-metal atoms, but this is not always the case.

DATIVE COVALENT BONDS

DON'T FORGET

A dative covalent bond is identical to other covalent bonds once it has been formed.

When a covalent bond is formed, two atomic orbitals combine together to form a molecular orbital. Usually, both atomic orbitals are half-filled before they merge to form the molecular orbital. However, a **dative covalent bond** is formed when **one atom provides both the electrons** that form the bond. The dative covalent bond is exactly the same as any other covalent bond after it has been formed – the only difference is in its formation.

An example of the formation of a dative covalent bond is when an ammonia molecule picks up a hydrogen ion to form an ammonium ion in solution. A hydrogen ion has no electrons and so cannot contribute an electron to the bond. Both the electrons come from the lone pair on the nitrogen atom in the ammonia molecule as shown on page 15.

LEWIS ELECTRON DOT DIAGRAMS

Lewis dot, or dot and cross, diagrams are used to represent bonding and non-bonding electron pairs in molecules and polyatomic ions.

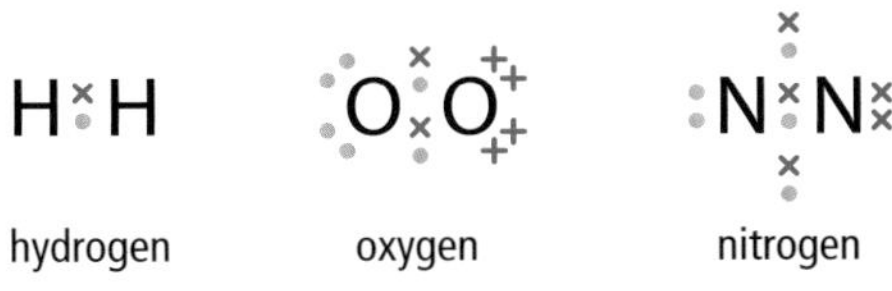

You can use dots to represent all the electrons, or you can use dots to represent only the electrons from one atom and crosses to represent electrons from the other atom. The diagrams above show that there is a single bond between the hydrogen atoms, a double bond (four electrons) between the oxygen atoms and a triple bond (six electrons) between the nitrogen atoms.

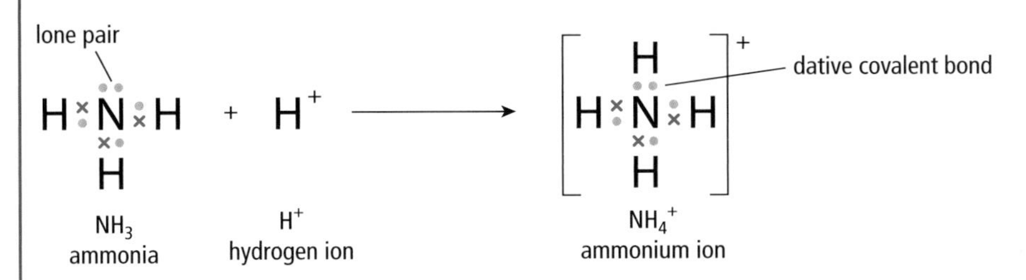

A Lewis dot and cross diagram can be used to show the formation of the dative covalent bond when an ammonia molecule and a hydrogen ion combine together to form the ammonium ion.

DON'T FORGET

The electron dot diagrams must show **all the outer electrons** of the atoms involved. Electrons in inner shells can be ignored.

LET'S THINK ABOUT THIS

1 Let us look at what happens when two hydrogen atoms approach one another and their atomic orbitals merge to form a molecular orbital; a covalent bond is formed between the two atoms.

At the start (at the right-hand side of the diagram) the two hydrogen atoms are far apart and are not interacting. This is position A on the graph. Now try to explain why the potential energy drops as the hydrogen atoms approach each other (positions B and C on the graph). Why is the potential energy at its lowest at point D? Why does the potential energy rise as the atoms get even closer together (point E on the graph)?

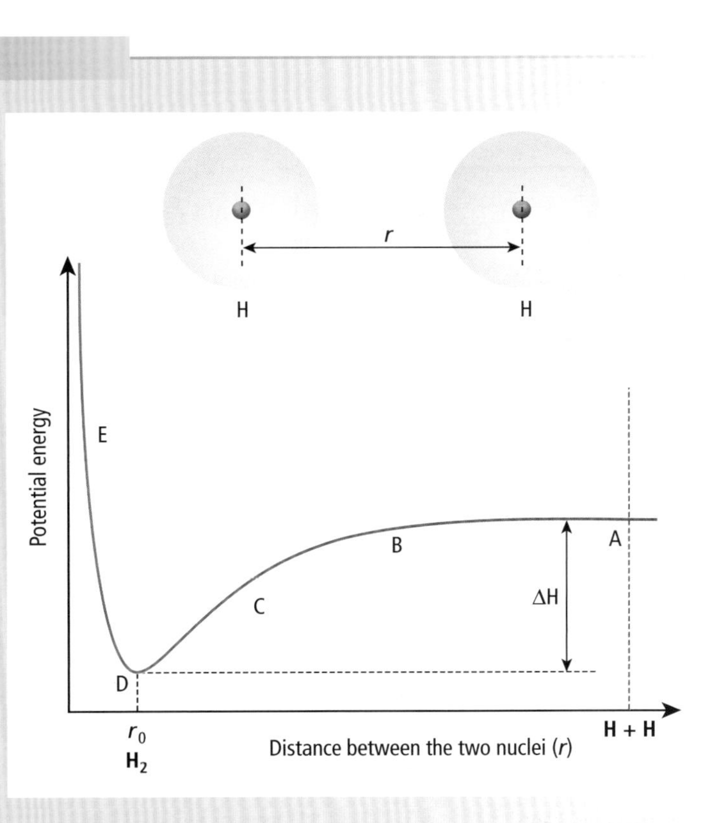

2 Both water and sodium hydride have $\Delta EN = 1{\cdot}3$ between their elements. However they have different types of bonding. Water has polar covalent bonding and sodium hydride is ionic. How would you show that sodium hydride has ionic bonding?

Websites that give you more information about dative covalent bonding include:
http://www.chemguide.co.uk/atoms/bonding/dative.html
http://www.webchem.net/notes/chemical_bonding/dative_bonding.htm

COVALENT BONDING, SHAPES OF MOLECULES AND POLYATOMIC IONS

RESONANCE STRUCTURES

If you draw the Lewis electron dot diagram for ozone, O_3, you start by working out that there are six outer electrons from each oxygen atom, giving a total of 18 electrons to be accommodated.

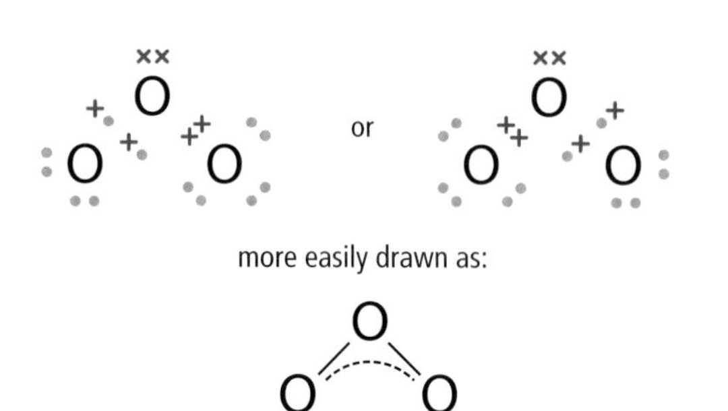

*Note that each oxygen atom is stable since it has a share of **eight** outer electrons. The oxygen at the top in each representation has two of its own electrons plus a half-share of four of its own electrons. It also has a half-share of another two electrons. Effectively, it has lost an electron so should have a positive charge. On the left-hand diagram, the oxygen on the right has effectively gained an electron and so should have a negative charge. On the right-hand diagram, the oxygen on the left has effectively gained an electron and should have a negative charge.*

The form that the molecule or ion adopts is the 'average' (or hybrid) of the resonance forms.

The two ozone dot and cross diagrams at the top are known as **resonance** structures. Double bonds (those with four electrons being shared) are shorter than single bonds (two shared electrons). However, the bonds in ozone are exactly the same length and a better representation is the diagram below the two resonance structures. The actual ozone molecule is a **hybrid** of the two resonance structures. The oxygen at the top may be shown with a positive charge and the negative charge shared between, or delocalised between, the other two oxygen atoms.

Important points to remember about these resonance forms are:

- The molecule is **not** rapidly changing between the different resonance structures. There is only **one** type of ozone molecule. (Think of a mule as a hybrid between a horse and a donkey. It is not a horse 50% of the time and a donkey the other 50% – it is a mule all the time!)
- The bond lengths between the oxygen atoms in ozone are **intermediate** between typical single O–O and double O=O bond lengths.
- We draw two Lewis structures for ozone because a single structure is insufficient to describe the real structure.

You can find out more about resonance structures (including those for the nitrate ion) at http://www.mikeblaber.org/oldwine/chm1045/notes/Bonding/Resonan/Bond07.htm or http://www.nku.edu/~russellk/tutorial/reson/resonance.html

SHAPES OF MOLECULES AND POLYATOMIC IONS

DON'T FORGET

If all the electron pairs are bonding pairs, the arrangement of the bonds will be the same as is given in the table and the molecules will have this shape.

Shapes of molecules and polyatomic ions such as NH_4^+ can be predicted by first calculating the total number of outer electron pairs around the central atom and then dividing them into bonding pairs and non-bonding (or lone) pairs.

The shape taken up by the molecule or polyatomic ion is that in which these electron pairs can be as far apart as possible. This arrangement minimises repulsion between the electron pairs. As shown in the table, the shape formed by the electron pairs depends on the total number of electron pairs.

Total number of electron pairs	Arrangement of electron pairs
2	linear
3	trigonal
4	tetrahedral
5	trigonal bipyramidal
6	octahedral

contd

SHAPES OF MOLECULES AND POLYATOMIC IONS contd

Some examples of molecules with different shapes

Two bonding pairs, for example $BeCl_2$(g)

Beryllium is in Group 2 and so has two outer electrons. The two Cl atoms contribute one electron each. This gives four electrons in two electron pairs. Since $BeCl_2$ has two Be–Cl bonds, the two electron pairs are two bonding pairs; there are two bonds around the central Be atom. Thus, beryllium chloride will be a linear molecule, Cl–Be–Cl with bond angle equal to 180°.

Three bonding pairs, for example BCl_3(g)

Boron is in Group 3 and so has three electrons in the outer shell. The three Cl atoms contribute one electron each, giving a total of six electrons involved in bonding. So, there are three B–Cl bonds and no lone pairs on the boron. The shape of the BCl_3 molecule will be trigonal (or trigonal planar) with all four atoms in the same plane.

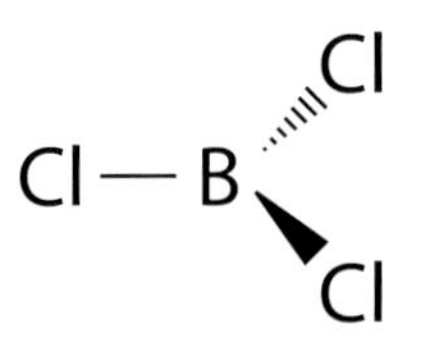

Boron chloride is trigonal planar.

Four bonding pairs, for example CH_4

Methane is a perfect tetrahedron with bond angles of 109·5°.

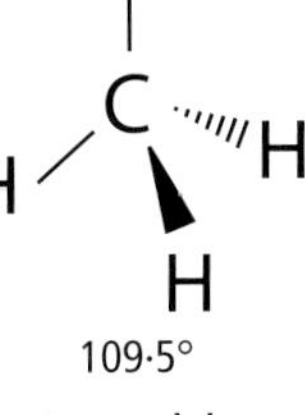

In methane the angle between the bonds is 109·5°. The molecule is a perfect tetrahedron.

Four pairs of electrons (three bonding and one lone pair), for example NH_3

Nitrogen has five outer electrons and each hydrogen atom contributes one electron, giving a total of eight electrons (in four pairs) around the N atom. So, in ammonia there are three N–H bonds and a non-bonding pair of electrons known as a **lone pair**. Repulsion between lone pairs and bonding pairs is greater than between different bonding pairs. The lone pair on the N atom 'squeezes' the three N–H bonds slightly closer together giving bond angles in NH_3 of 107°, which is slightly less than that in a true tetrahedron.

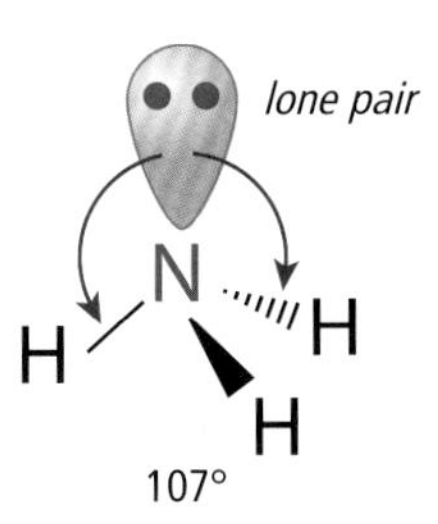

The bond angles in ammonia are reduced to 107° by the action of a lone electron pair.

Four pairs of electrons (two bonding pairs and two lone pairs), for example H_2O

In water, oxygen has six outer electrons and each hydrogen atom contributes one electron, giving a total of eight electrons around the O atom. There are four electron pairs, but only two O–H bonds; the other electron pairs will be non-bonding or lone pairs. The two lone pairs around the O atom squeeze the two O–H bonds closer together, and so the bond angle in H_2O is approximately 104·5°.

O
H H
104·5°

In water the action of the two lone pairs reduces the bond angle to 104·5°.

Five bonding pairs, for example PCl_5(g)

Phosphorus has five outer electrons and each of the five chlorine atoms contribute one electron, giving a total of 10 electrons (five electron pairs). All the electron pairs are P–Cl bonds. The shape of the molecule is trigonal bipyramidal.

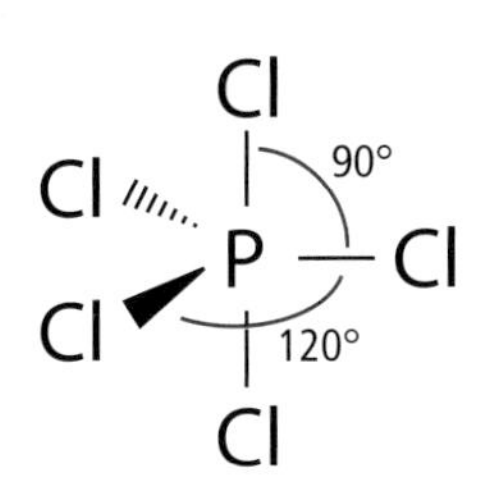

Phosphoros pentachloride is trigonal bypyramidal.

Six bonding pairs, for example, SF_6(g)

Sulphur has six outer electrons and each of the six fluorine atoms contribute one electron, giving a total of 12 electrons in six electron pairs. All the electron pairs are S–F bonds, so the shape of the molecule is octahedral.

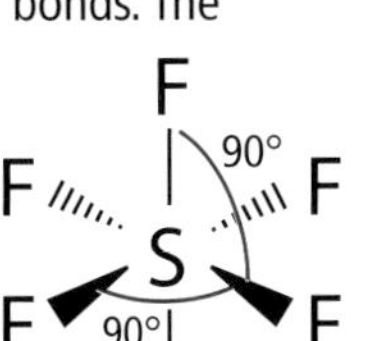
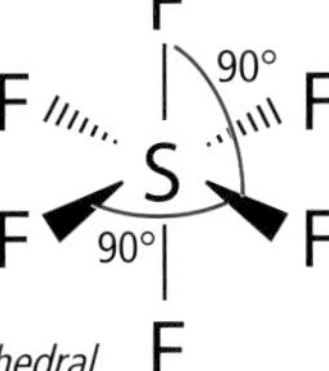

Sulphur hexafluoride is octahedral.

LET'S THINK ABOUT THIS

1 Draw resonance structures for **(a)** sulphur dioxide and **(b)** the ethanoate ion.

2 Calculate the number of bonding and non-bonding electron pairs around the central atom in the following species and work out the molecular shape:
(a) SiF_4 **(b)** PCl_3 **(c)** Cl_2O **(d)** ClF_3 **(e)** I_3^-

Websites where you can get more information include
http://intro.chem.okstate.edu/1314f00/lecture/chapter10/vsepr.html
http://www.chemguide.co.uk/atoms/bonding/shapes.html
http://www.chemmybear.com/shapes.html

IONIC LATTICES, SUPERCONDUCTORS AND SEMICONDUCTORS

IONIC LATTICE STRUCTURES

DON'T FORGET

There are two types of ionic lattice structure that you have to know. These are the sodium chloride-type structure, which many ionic compounds adopt, and the caesium chloride-type structure, adopted by fewer ionic compounds.

As you know, ionic compounds have high melting and boiling points and are solid at room temperature. Ionic compounds only conduct electricity when in solution or when molten since, under these conditions, the ions are free to move. Ionic compounds do not conduct electricity in the solid state when the ions are trapped in a structure known as a lattice. The positive and negative ions are held in the lattice by electrostatic attractions between their opposite charges.

The type of lattice structure which an ionic compound adopts depends upon the relative sizes of the ions. The actual structure will be the one which is the best compromise between the minimum repulsion of similarly charged ions and the maximum attraction of oppositely charged ions. In general, if the ions are about the same size, then the ionic lattice adopted is the caesium chloride structure. If one ion is much bigger than the other, the sodium chloride structure is more likely.

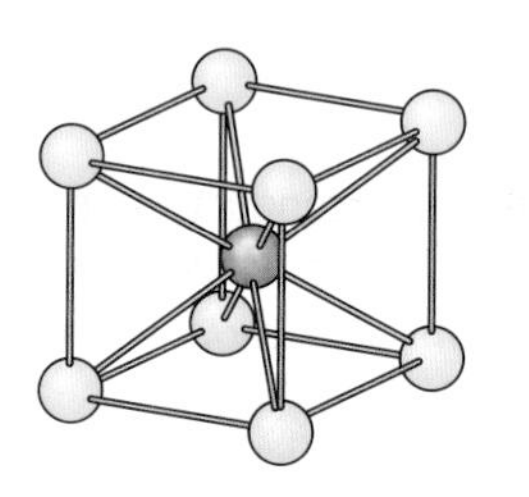

The caesium chloride structure is sometimes referred to as a simple cubic lattice.

Caesium ions, Cs^+, are much bigger than sodium ions, Na^+, and one caesium ion can accommodate eight negative chloride ions around it, without the negative ions being too close together. Each chloride ion is also surrounded by eight caesium ions. We say that this type of lattice structure has 8:8 coordination or that the coordination number is 8.

The ionic radii of Cs^+ and Cl^- given in the SQA Data Booklet are 174 pm and 181 pm respectively and so the radius ratio is approximately = 1.

DON'T FORGET

The NaCl structure has 6:6 coordination and the CsCl structure has 8:8 coordination.

The ionic radius of Na^+ is 95 pm, which is much smaller than a chloride ion. Only six chloride ions can be accommodated around a smaller sodium ion without the repulsive effect of the chloride ions being greater than the attractive forces between the oppositely charged ions. The sodium chloride lattice structure has 6:6 coordination or the coordination number = 6.

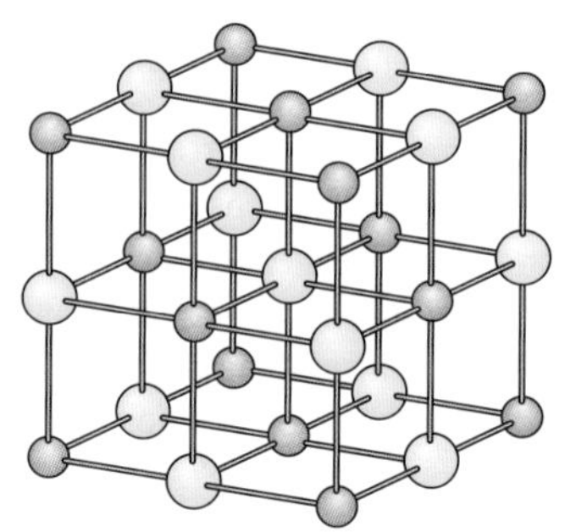

The sodium chloride structure is sometimes referred to as a simple cubic lattice.

Most ionic compounds adopt either the sodium chloride-type or the caesium chloride-type lattice structure. You can predict the type of structure by calculating the radius ratio, that is by dividing the radius of the positive ion by the radius of the negative ion. If the value is close to 1, then the material will take on the caesium chloride-type structure. If it is closer to 0·5, then the sodium chloride-type structure is more likely.

SUPERCONDUCTORS

Superconductors are a special class of materials that have zero electrical resistance. Unfortunately, this is usually only true at temperatures close to absolute zero (0 K or –273 °C) when liquid helium must be used as the coolant. Since it is impractical and costly to maintain a temperature as low as this, research is being carried out to develop materials which act as superconductors at higher temperatures.

The boiling point of nitrogen is –196°C and some materials will act as superconductors when immersed in liquid nitrogen. It is much cheaper to use liquid nitrogen than liquid helium. The critical temperature (T_c), of a superconductor is the temperature below which it has zero electrical resistance. The compound $YBa_2Cu_3O_7$, sometimes known as the 1–2–3 superconductor has a critical temperature of about 92 K (–181 °C).

contd

SUPERCONDUCTORS contd

Superconductors may have future applications in reducing energy loss during power transmission and in electrically powered forms of transport which would be almost frictionless. Superconductors are already in use in magnetic resonance imaging (MRI) in medicine.

SEMICONDUCTORS

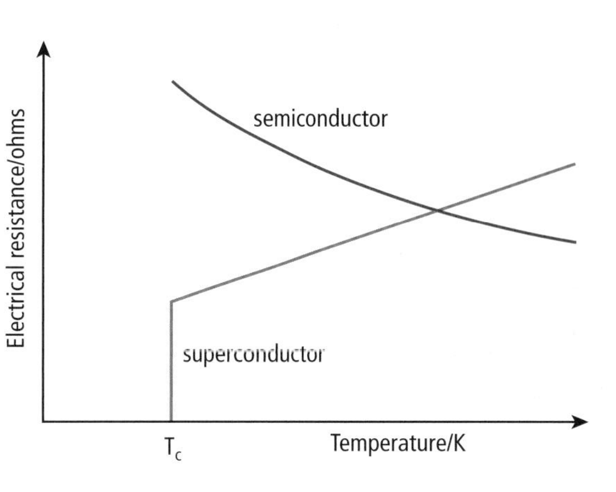

Graphite is a very good electrical conductor because, like a metal, it has delocalised electrons. Most covalent elements are considered non-conductors. However, some covalent elements such as silicon and germanium have higher electrical conductivity than typical non-metals but much lower conductivity than metals. As such, they are known as semiconductors. Semiconductors are sometimes referred to as metalloids and are found at the division between metal and non-metal elements in the Periodic Table.

Unlike superconductors and metals, the **electrical resistance** of semiconductors decreases as the temperature increases, and so the **electrical conductivity** of semiconductors increases as the temperature increases. The electrical conductivity of semiconductors also increases on exposure to light. This is known as the photoelectric effect. The light energy is picked up by electrons in the atoms of the substance, allowing them to move around, and so produce the current.

Although the bonding structures in silicon and germanium are similar to diamond, the covalent bonds are weaker. This results in a conductivity that is higher than the conductivity of a non-metal such as sulphur which has no free electrons, but lower than that of a metal which has mobile electrons. Increasing the temperature can cause some of the bonding electrons to break free, leaving sites known as **positive holes**. When a voltage is then applied to these elements, electrons and positive holes can migrate through the lattice. The movement of electrons in one direction is equivalent to the movement of positive holes in the other direction. This gives the element a low conductivity – it is a semiconductor.

Conduction can be considerably affected by the presence of impurities. Deliberate addition of impurities to improve conductivity is called **doping**. Doping pure crystals of silicon or germanium with a Group 5 element such as arsenic or phosphorus introduces an extra electron into the lattice structure and an **n-type semiconductor** is formed. In an **n**-type semiconductor the main current carrier is surplus **n**egative electrons.

If the dopant is a Group 3 element such as boron or aluminium then, because these elements have only three outer electrons (one less than silicon and germanium), **p**ositive holes are the main current carrier; a **p-type semiconductor** is made.

A **p–n junction** is formed between a layer of n-type and a layer of p-type semiconductor. This is used in photovoltaic cells, particularly solar cells. In a photovoltaic cell, photons of light energy excite electrons so that they can move freely in the semiconductor. The positions left free by the electrons become positive holes. The excited electrons and positive holes move in opposite directions and so sunlight can be converted into electricity without any mechanical moving parts being involved.

DON'T FORGET

As temperature increases, the conductivity of semiconductors increases but superconductors only have zero electrical resistance **below a certain temperature**.

You will find more information about the photoelectric effect at http://www.einsteinyear.org/facts/photoelectric_effect/

You can get more information about photovoltaic cells at http://www.cea.fr/var/cea/storage/static/gb/library/Clefs50/pdf/encadred.pdf

LET'S THINK ABOUT THIS

1 Draw a graph of conductivity versus temperature for a semiconductor and for a superconductor.

2 Using the ionic radius values given on page 16 of the SQA Data Booklet, work out the likely ionic lattice structure for **(a)** Li^+F^- **(b)** $Mg^{2+}O^{2-}$ **(c)** $Ba^{2+}O^{2-}$.

OXIDES, CHLORIDES AND HYDRIDES

OXIDES

Oxygen is in Group 6 of the Periodic Table and, with an electronegativity value of 3·5, it is one of the most electronegative elements. As a result, it combines readily with most other elements. In general, oxides of metallic elements are usually ionic and basic; they will neutralise acids to form a salt and water. Oxides of non-metallic elements are usually covalent and acidic, some dissolving in water to form acidic solutions.

The bonding structure and properties of the **simplest oxides** of the elements in the third period can be seen in the table.

Group	1	2	3	4	5	6	7
Formula	Na_2O	MgO	Al_2O_3	SiO_2	P_2O_5	SO_2	Cl_2O
Bonding structure	ionic lattice	ionic lattice	ionic lattice	covalent network	covalent molecular	covalent molecular	covalent molecular
Acid/base character	basic	weakly basic	amphoteric	acidic	acidic	acidic	acidic

DON'T FORGET

It is important to know that SiO_2 has a covalent network structure similar to diamond and has very high melting and boiling points.

As the electronegativity difference between oxygen and the other element in the oxide decreases, the bonding becomes less ionic and more covalent.

Aluminium oxide

Aluminium oxide is an amphoteric oxide. Amphoteric oxides have both acidic and basic properties. For example, aluminium oxide will neutralise both acids and alkalis to form a salt and water, as shown in these equations:

$Al_2O_3 + 6HCl \rightarrow 2AlCl_3 + 3H_2O$

$Al_2O_3 + 2NaOH \rightarrow 2NaAlO_2 + H_2O$

The salt formed here is known as sodium aluminate. Other amphoteric oxides include BeO and H_2O.

DON'T FORGET

Amphoteric oxides exhibit both acidic and basic properties. Aluminium oxide is an amphoteric oxide.

CHLORIDES

Chlorine is in Group 7 of the Periodic Table and, with an electronegativity value of 3·0, it is another one of the most electronegative elements.

Most metal chlorides have ionic bonding and dissolve in water without reacting. However, some covalent chlorides are easily hydrolysed in water, or even in moist air, to produce white fumes of hydrogen chloride gas. For example, the equation for the reaction of silicon chloride and water is:

$SiCl_4 + 2H_2O \rightarrow SiO_2 + 4HCl$

The bonding structure and properties of the **simplest chlorides** of the elements in the third period can be seen in the table.

Group	1	2	3	4	5	6	7
Formula	$NaCl$	$MgCl_2$	$AlCl_3$	$SiCl_4$	PCl_3	SCl_2	Cl_2
Bonding structure	ionic lattice	ionic lattice	covalent molecular	covalent molecular	covalent molecular	covalent molecular	covalent molecular
'Reaction' with water	soluble	soluble	fumes of HCl produced	fumes of HCl produced	fumes of HCl produced	fumes of HCl produced	dissolves to form acidic solution

DON'T FORGET

You should know the covalent chlorides which form white fumes of HCl when in contact with water.

contd

CHLORIDES contd

Aluminium chloride

Despite being the chloride of a metal element, aluminium chloride has mainly covalent character and is quickly hydrolysed by water, producing the characteristic white fumes of HCl. Solid aluminium chloride sublimes (changes directly from the solid state to the gaseous state) and in the gaseous state it has molecular formula Al_2Cl_6, i.e. it is a dimer.

Carbon tetrachloride

You should also know that CCl_4 is a covalent chloride with four polar C–Cl bonds. However, because of its tetrahedral molecular shape and its symmetry, it is non-polar overall. It is immiscible in water and does **not** react with water to form white fumes of HCl.

You can find more information about the chlorides at http://www.chemguide.co.uk/inorganic/period3/chlorides.html

HYDRIDES

Hydrogen has electronegativity value of 2·2 which means it is neither one of the most, nor one of the least electronegative elements.

Most metal hydrides have ionic bonding and contain the hydride ion, H^-. They react with water to produce hydrogen gas and leave behind an alkaline solution. The equation for sodium hydride reacting with water is:

$Na^+H^-(s) + H_2O(l) \rightarrow H_2(g) + Na^+OH^-(aq)$

The hydride ion can act as a reducing agent. In organic chemistry it is used in the form of lithium aluminium hydride, $LiAlH_4$, or sodium borohydride, $NaBH_4$, to reduce aldehydes to primary alcohols.

The bonding structure and properties of the **simplest hydrides** of the elements in the third period can be seen in the table.

Group	1	2	3	4	5	6	7
Formula	NaH	MgH_2	AlH_3	SiH_4	PH_3	H_2S	HCl
Bonding structure	ionic lattice	intermediate between ionic and covalent	polymeric	covalent molecular	covalent molecular	covalent molecular	covalent molecular
Acid/base character	strongly alkaline	alkaline	alkaline	neutral	insoluble in water	weakly acidic	strongly acidic

Electrolysis of molten ionic hydrides produces hydrogen gas at the **positive** electrode. The ion-electron equation is:

$2H^-(l) \rightarrow H_2(g) + 2e^-$

Ionic hydrides are being considered as a possible means of storing hydrogen in hydrogen-powered vehicles.

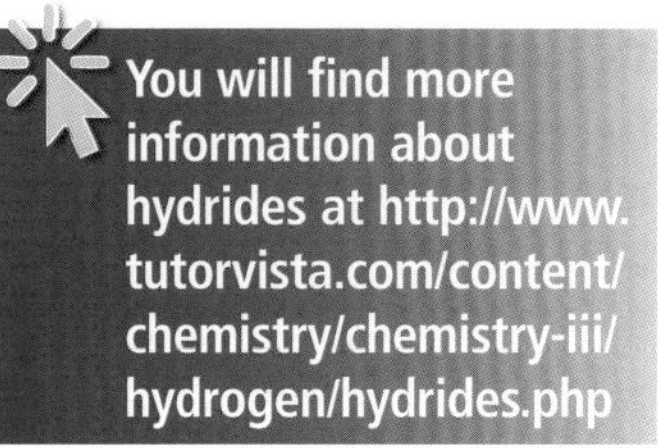

LET'S THINK ABOUT THIS

Here are some questions from recent Advanced Higher exam papers. You should be able to answer them now.

1 Which of the following is the best term to describe aluminium oxide?
A Basic **B** Acidic **C** Neutral **D** Amphoteric

2 Which of the following compounds contains hydride ions?
A NH_3 **B** HCl **C** H_2S **D** CaH_2

3 Which of the following hydrides, when added to water, would give the most acidic solution?
A Sodium hydride **B** Magnesium hydride **C** Silicon hydride **D** Sulphur hydride

4 Which of the following compounds would produce fumes of hydrogen chloride when added to water?
A LiCl **B** $MgCl_2$ **C** PCl_3 **D** CCl_4

5 Which of the following compounds shows most covalent character?
A CH_4 **B** NaH **C** NH_3 **D** PH_3

ELECTRONIC CONFIGURATION AND OXIDATION STATES OF TRANSITION METALS

ELECTRONIC CONFIGURATION

The d block transition metals are metals with an incomplete d subshell in at least one of their ions. Normally we would consider the first row of the transition metals as being from scandium to zinc and the second row from ytrrium to cadmium. Platinum and gold are in the third row of transition elements. Most of the common metals in everyday use are transition metals.

When we consider the electronic configurations of the elements from zinc to copper we are, in the main, filling the 3d subshell according to the aufbau principle. Once again, the electronic configuration has to fit in with the electron arrangement given in the SQA Data Booklet.

The table below shows the electronic configuration in spectroscopic notation and in orbital box notation for the elements from scandium to zinc. [Ar] represents the electronic configuration of argon which is $1s^2\ 2s^2\ 2p^6\ 3s^2\ 3p^6$. It is okay to use this shorthand here, instead of writing out the full electron shells up to 3p. However, in an exam you should write out the spectroscopic notation for each element in full.

	Electronic configuration	
Element	**Spectrosopic notation**	**Orbital box notation (d electrons only)**
scandium	$[Ar]\ 3d^1\ 4s^2$	[↑] [ ] [ ] [ ] [ ]
titanium	$[Ar]\ 3d^2\ 4s^2$	[↑] [↑] [ ] [ ] [ ]
vanadium	$[Ar]\ 3d^3\ 4s^2$	[↑] [↑] [↑] [ ] [ ]
chromium	$[Ar]\ 3d^5\ 4s^1$	[↑] [↑] [↑] [↑] [↑]
manganese	$[Ar]\ 3d^5\ 4s^2$	[↑] [↑] [↑] [↑] [↑]
iron	$[Ar]\ 3d^6\ 4s^2$	[↑↓] [↑] [↑] [↑] [↑]
cobalt	$[Ar]\ 3d^7\ 4s^2$	[↑↓] [↑↓] [↑] [↑] [↑]
nickel	$[Ar]\ 3d^8\ 4s^2$	[↑↓] [↑↓] [↑↓] [↑] [↑]
copper	$[Ar]\ 3d^{10}\ 4s^1$	[↑↓] [↑↓] [↑↓] [↑↓] [↑↓]
zinc	$[Ar]\ 3d^{10}\ 4s^2$	[↑↓] [↑↓] [↑↓] [↑↓] [↑↓]

DON'T FORGET

Electronic configurations must fit in with electron arrangements given in the SQA Data Booklet. Remember that the electronic configurations of Cr and Cu are exceptions to the aufbau principle.

Chromium and copper

You can see from the electron arrangements in the SQA Data Booklet and the electronic configurations written in spectroscopic notation in the table that chromium and copper seem to be out of step with the aufbau principle. However, there is a special stability associated with half filled or completely filled d orbitals. Bear this in mind when looking at the orbital box notation and you can understand why Cr is $[Ar]\ 3d^5\ 4s^1$ and Cu is $[Ar]\ 3d^{10}\ 4s^1$, rather than the $[Ar]\ 3d^4\ 4s^2$ and $[Ar]\ 3d^9\ 4s^2$ which you might have expected.

However, when any transition metal atom forms an ion the **electrons that are lost first** are those in the outer subshell, the **4s electrons**. Therefore, the electronic configuration of the Co^{2+} ion is $[Ar]\ 3d^7$.

DON'T FORGET

When a transition metal atom forms an ion, the 4s electrons are lost before any 3d electrons.

OXIDATION STATES

The oxidation state is similar to the valency that an element has when it is part of a compound. For example, in iron(II) chloride we might say that the iron has valency 2. However, it is actually more accurate to say that iron is in oxidation state (II) or has oxidation number +2.

There are certain rules to be followed when assigning an oxidation number to an element:

- The oxidation number of an uncombined element is 0. So, elemental Na has oxidation number 0, as does F in fluorine gas, F_2.
- For ions containing single atoms (monatomic ions) such as Na^+, Cl^- or O^{2-}, the oxidation number is the same as the charge on the ion. In the examples above, this would be +1, –1 and –2.
- In most of its **compounds**, oxygen has oxidation number –2.
- In most of its **compounds,** hydrogen has oxidation number +1. The exception is in metallic hydrides when hydrogen has oxidation number –1.
- Fluorine always has oxidation number –1 in all its **compounds**.
- The sum of all the oxidation numbers of all the atoms in a molecule or neutral compound must add up to zero.
- The sum of all the oxidation numbers of all the atoms in a polyatomic ion must add up to the charge on the ion.

Calculating an oxidation state

For example, if we want to find the oxidation number of Mn in MnO_4^-, the sum of the oxidation numbers of the one Mn atom and the four oxygen atoms must add up to –1, since this is the charge on the ion. Each oxygen atom has oxidation number –2, and so sum of the oxidation numbers of the four oxygen atoms must be –8. Therefore, the oxidation number of Mn must be 7, since $7 - 8 = -1$. We can say the manganese has oxidation number +7 or is in oxidation state (VII).

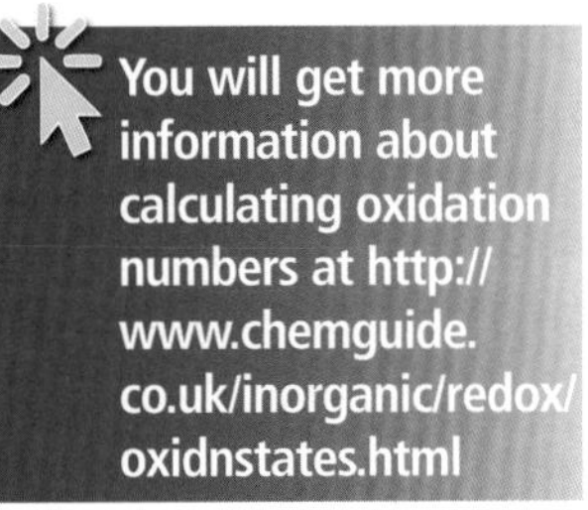

Multiple oxidation states

The same transition metal may have different oxidation states in its compounds, for example you know that the common oxidation states of iron are (II) and (III). In its compounds, copper is usually in oxidation state (II) but it can also have oxidation number +1 in, for example, Cu_2O.

Iron(III) is usually more stable than iron(II); iron(II) compounds in solution are often unstable since they slowly become oxidised to iron(III) compounds. Sometimes, transition metal compounds have different colours depending on the oxidation state of the metal. For example, iron(II) compounds are often a pale green colour which slowly changes to the familiar yellow–orange colour of iron(III) compounds as oxidation occurs.

The ion-electron equation for iron(II) ions changing to iron(III) ions is: $Fe^{2+} \rightarrow Fe^{3+} + e^-$

This is a loss of an electron and so represents oxidation.

So, **oxidation** can be redefined as an **increase in oxidation number**. **Reduction** can be redefined as a **decrease in oxidation number**.

Oxidising and reducing agents

Acidified permanganate is a very good oxidising agent and the relevant ion-electron equation for this is: $MnO_4^-(aq) + 8H^+(aq) + 5e^- \rightarrow Mn^{2+}(aq) + 4H_2O(l)$

The Mn is changing from oxidation state (VII) to oxidation state (II). In general, compounds containing metals in high oxidation states tend to be good oxidising agents, since the ions are easily reduced to lower oxidation states. Likewise, compounds containing metals in low oxidation states tend to be reducing agents.

Oxidation can be considered an increase in oxidation number and reduction a decrease in oxidation number.

LET'S THINK ABOUT THIS

1. The d block transition metals are metals with an incomplete d-subshell in at least one of their ions. Try to explain why Sc and Zn are often considered **not** to be transition metals.
2. Consider the electronic configurations of the Fe^{2+} and Fe^{3+} ions in both spectroscopic and orbital box notations. Use these notations to explain why Fe(III) compounds are more stable than Fe(II) compounds.
3. Work out the oxidation number of Cr in $Cr_2O_7^{2-}(aq)$ and explain why acidified dichromate is a good oxidising agent.

TRANSITION METAL COMPLEXES 1

LIGANDS AND COMPLEXES

DON'T FORGET

Ligands must have at least one lone pair of electrons and these are shown as a pair of dots on the atoms.

A **complex** consists of a central metal ion surrounded by ligands. **Ligands** are negative ions or uncharged molecules with one or more lone pairs of electrons. They are electron donors, donating these non-bonding electrons into unfilled metal orbitals thereby forming dative covalent bonds.

Ammonia and water are very common neutral ligands. Ammonia has a lone pair of electrons on the N atom and water has two lone pairs on the O atom. Common negatively charged ligands include Cl^- and the cyanide ion, CN^-.

Ligands which donate **one** pair of electrons to the central metal ion are said to be **monodentate**. Examples include Cl^- and H_2O. A **bidentate** ligand donates **two** pairs of electrons to the central metal ion. The Unit 1 PPA 1 experiment in which potassium trioxalatoferrate(III) is prepared involves the oxalate ion as a ligand (see page 28).

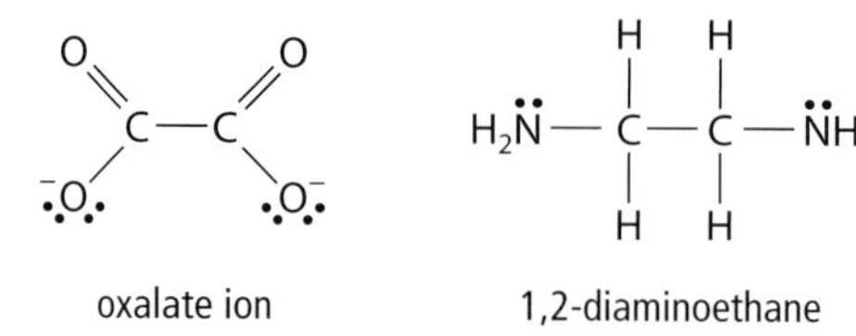

Examples of bidentate ligands include the oxalate ion, $C_2O_4^{2-}$, and 1,2-diaminoethane. These have lone pairs of electrons on the oxygen and nitrogen atoms respectively.

A ligand that you use in Unit 2 PPA 1 is ethylenediaminetetraacetic (EDTA) (see page 52). This is a hexadentate ligand since it has six pairs of non-bonding electrons which bind to a central metal ion forming a complex. It reacts with metal ions such as Ni^{2+} in a 1 : 1 ratio.

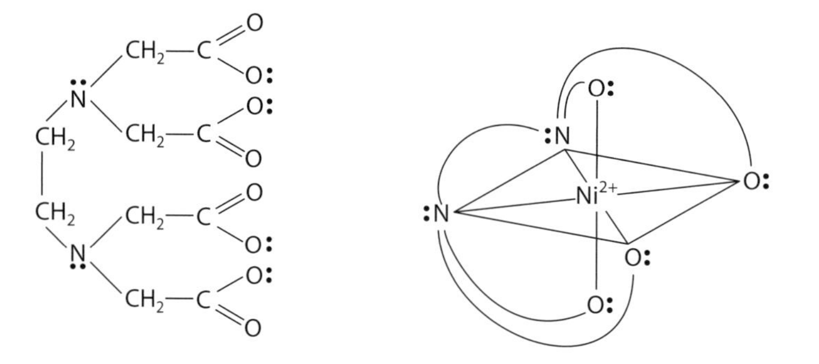

The number of bonds from the ligand to the central metal ion is known as the ***coordination number*** *of the central metal ion. In the Ni^{2+}/EDTA complex above, the Ni^{2+} ion has coordination number = 6.*

NAMING COMPLEXES

Complexes and complex ions are named and written according to IUPAC rules – the International Union of Pure and Applied Chemistry is the world authority on chemical nomenclature.

When writing the formula of a complex, the symbol of the metal is written first, then the negatively charged ligands, followed by the neutral ligands. Finally, the formula of the complex ion is enclosed within square brackets as in, for example, $[Fe(OH)_2(H_2O)_6]^+$. This ion has overall charge +1 as this is the sum of the Fe^{3+} ion and the two OH^- ions.

Rules for naming complexes are:

- The ligands are named first in alphabetical order, followed by the name of the metal and its oxidation state.
- If the ligand is a negative ion ending in –ide, then in the complex the ligand name changes to end in 'o'. Examples are chloride which becomes 'chloro' and cyanide which becomes 'cyano'.

contd

NAMING COMPLEXES contd

- If the ligand is ammonia, NH_3, it is named 'ammine'. Water as a ligand is called 'aqua'.
- If the complex is a negative ion overall, the name of the complex ion ends in –ate. An example is cobaltate for a negative ion containing cobalt. However, cuprate is used if the complex contains copper and ferrate is used if it contains iron, from their Latin names (not copperate or ironate).

DON'T FORGET

When ammonia is a ligand it is named ammine and water is named aqua. It is important that these are spelled correctly.

Examples of complex names

- The complex ion $[Cu(H_2O)_4]^{2+}$ is named the tetraaquacopper(II) ion.
- $[Co(NH_3)_6]^{2+}$ is named hexaamminecobalt(II).
- $[Fe(CN)_6]^{2-}$ is named hexacyanoferrate(II) since it is a negative ion and the iron is in oxidation state (II).

LOSS OF DEGENERACY AND THE SPECTROCHEMICAL SERIES

In a free transition metal or ion (one which is not complexed to any ligands) the five different d orbitals in the 3d subshell are degenerate.

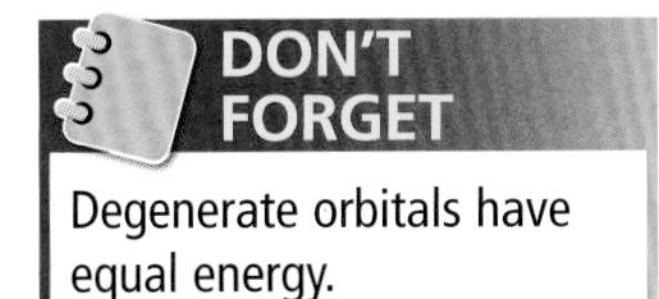

DON'T FORGET

Degenerate orbitals have equal energy.

Now consider the formation of an octahedral complex such as $[Ni(H_2O)_6]^{2+}$. Think of six water ligands approaching the Ni^{2+} ion along the x-, y- and z-axes. The electrons in the nickel ion's d orbitals that lie along the axes will be repelled by the electrons of the approaching ligands.

As a result, these d orbitals now have higher energy than the d orbitals that lie between the axes. So, the d orbitals are no longer degenerate. The d orbitals which lie on the axes are the $d_{x^2-y^2}$ (a double dumbbell lying on both the x-and y-axes) and the d_{z^2} which lies on the z-axis. The lower energy orbitals are the d_{xy}, d_{xz}, and d_{yz} orbitals (double dumbbells which lie between the axes).

We call this 'splitting' of the d orbitals. The splitting is different in octahedral complexes compared to tetrahedral and other shapes of complexes.

DON'T FORGET

The energy difference depends on the position of the ligand in the spectrochemical series.

The energy difference between the different subsets of d orbitals depends on the position of the ligand in the **spectrochemical series**. This is a series which puts the ability of different ligands to split the d orbitals into order. Those ligands which cause a large energy difference in the d orbitals are said to be 'strong field' ligands, compared to 'weak field' ligands where the energy difference is small.

A short form of the spectrochemical series is $CN^- > NH_3 > H_2O > OH^- > F^- > Cl^- > Br^- > I^-$, in which the cyanide ion causes the greatest energy difference.

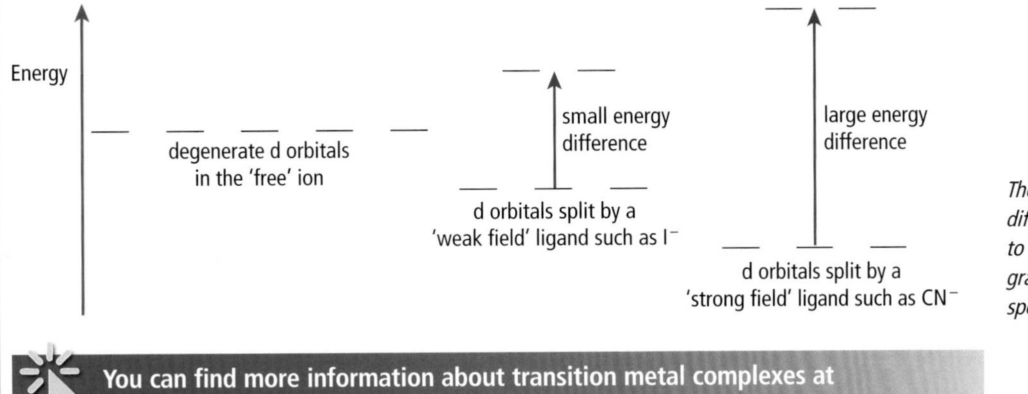

The diagram shows the differing abilities of ligands to split d orbitals; it is a graphical illustration of the spectrochemical series.

You can find more information about transition metal complexes at http://www.drbateman.net/asa2sums/sum5.2/sum5.2.htm

LET'S THINK ABOUT THIS

1 Name the following complex ions: **(a)** $[CoCl_4]^{2-}$ **(b)** $[Ni(NH_3)_6]^{2+}$ **(c)** $[PtCl_6]^{2-}$

TRANSITION METAL COMPLEXES 2

COLOUR IN TRANSITION METAL COMPOUNDS

Many transition metal compounds are coloured. For example, solutions of copper(II) compounds are usually blue and solutions of nickel(II) compounds are usually green. To explain how these colours arise we must examine simple colour theory.

- White light can be thought of as a combination of three primary colours – red, green and blue.
- If red light is absorbed, green and blue light are transmitted. We see this as a blue-green colour or cyan.
- If green light is absorbed, a combination of red and blue light is transmitted. We see this as purple or magenta.
- If blue light is absorbed, red and green light are transmitted which we see as yellow.

Why do transition metal compounds absorb light?

Think back to the split d orbitals. Electrons in the lower energy d orbitals can absorb energy and move to the higher energy d orbitals. If the energy absorbed in these so-called d–d transitions is in the visible part of the electromagnetic spectrum, the colour of the transition metal compound will be the complementary colour of the absorbed colour. So, the colour we see will be white light minus the colour absorbed.

UV AND VISIBLE SPECTROSCOPY

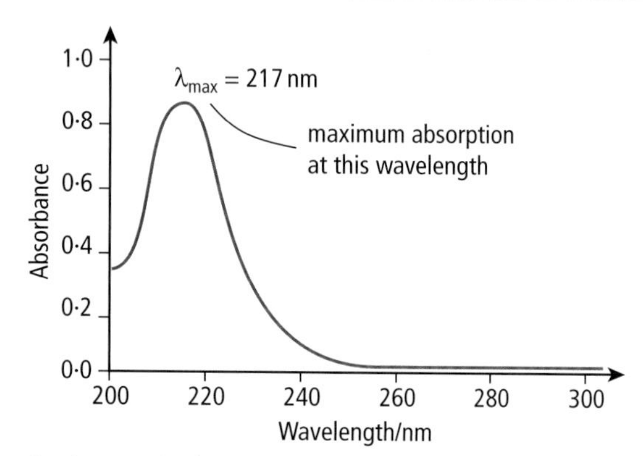

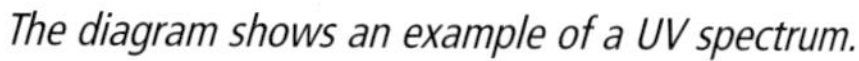
The diagram shows an example of a UV spectrum.

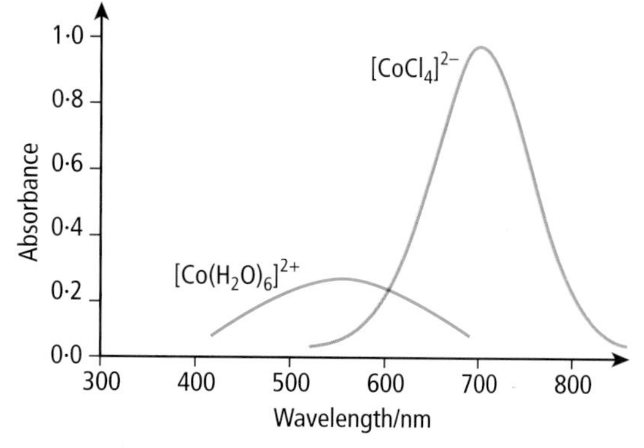

The diagram shows two examples of visible spectra.

The effects of d–d transitions can be studied using spectroscopy. If the absorbed energy is in the visible part of the electromagnetic spectrum giving a coloured compound, visible spectroscopy is used. If the absorbed energy is in the ultraviolet part of the electromagnetic spectrum, the compound will be colourless and UV spectroscopy is used.

- When the ligands surrounding the transition metal ion are strong-field ligands such as CN^-, d–d transitions are more likely to occur in the ultraviolet region. The wavelength range of ultraviolet light is approximately 200–400 nm.
- Complexes containing weak-field ligands such as H_2O are more likely to absorb visible light, making them coloured. The wavelength range for the visible region is approximately 400–700 nm.

A colorimeter fitted with coloured filters corresponding to certain wavelengths in the visible region can be used to measure the absorbance of coloured solutions. A filter of the complementary colour to the colour of the solution should be used. You use a colorimeter in the Unit 1 PPA 2 (see page 29).

A UV spectrometer is a bit more complicated than a colorimeter. Different wavelengths of ultraviolet light from 200–400 nm are passed through the sample and the quantity of UV light absorbed at different wavelengths is recorded. The results are graphed automatically as a UV spectrum. As with a colorimeter, the absorbance is directly proportional to the concentration of the absorbing species.

You will find more information about UV and visible spectroscopy at
http://www.chemguide.co.uk/analysis/uvvisible/analysis.html
http://www.chemguide.co.uk/analysis/uvvisible/theory.html
http://www.cem.msu.edu/~reusch/VirtualText/Spectrpy/UV-Vis/spectrum.htm

CATALYSIS

As you know, transition metals and their compounds are used as catalysts: for example, iron in the Haber Process; platinum in the Ostwald process; and nickel in the production of margarine. Platinum, rhodium and palladium are also used in the catalytic converters in the exhaust systems of cars with petrol engines. These are examples of **heterogeneous catalysts** since they are in a different physical state to the reactants in the reactions being catalysed.

It is thought that the transition metal atoms on the surface of the active sites can form weak bonds to the reactant molecules, probably using available partially filled or empty d orbitals to form intermediate complexes. The effect of this is to weaken the covalent bonds inside the reactant molecules and, since the reactant molecules are now also held in a favourable position, they are more susceptible to attack by the molecules of the other reactant. The overall effect is that an alternative pathway with lower activation energy is provided and the rate of the reaction increases.

Another reason transition metals may act as catalysts is because they have variable oxidation states. This also allows the transition metal to provide an alternative reaction pathway with lower activation energy, thus increasing the reaction rate. An example is vanadium(V) oxide which catalyses the reaction of sulphur dioxide with oxygen to form sulphur trioxide, eventually leading to the production of sulphuric acid.

Vanadium(V) oxide as a catalyst

The sulphur dioxide is oxidised to sulphur trioxide by the vanadium(V) oxide. In the process, the vanadium(V) oxide is reduced to vanadium(IV) oxide: $SO_2 + V_2O_5 \rightarrow SO_3 + V_2O_4$

The vanadium(IV) oxide is then re-oxidised by the oxygen: $V_2O_4 + \frac{1}{2}O_2 \rightarrow V_2O_5$

The overall reaction is: $SO_2 + \frac{1}{2}O_2 \rightarrow SO_3$

This is a good example of the way that a catalyst can be changed during the course of a reaction. At the end of the reaction, though, it will be chemically the same as it was at the start.

Cobalt(II) chloride as a catalyst

Another example is one you have covered at Higher Level – the reaction of a solution of Rochelle salt (potassium sodium tartrate) and hydrogen peroxide which can be catalysed by cobalt(II) chloride solution.

The cobalt(II) chloride solution is pink at the start, but during the reaction the pink colour changes to green as Co^{3+} ions form. At this stage there is vigorous effervescence as O_2 gas is given off. At the end of the reaction, Co^{2+} ions are regenerated and the reaction mixture returns to its original pink colour. This is an example of **homogeneous catalysis**.

$Co^{2+}(aq) \rightarrow Co^{3+}(aq) \rightarrow Co^{2+}(aq)$

pink green pink

You can find more information about transition metals acting as catalysts at http://www.chemguide.co.uk/physical/catalysis/introduction.html

LET'S THINK ABOUT THIS

1 Colours of transition metal compounds are usually explained in terms of d–d transitions, but this is not always the case. For example, the permanganate ion, MnO_4^-, has an intense purple colour. In this ion, manganese is in oxidation state (VII). The electronic configuration of elemental Mn is [Ar] $3d^5\ 4s^2$; the manganese ion in oxidation state (VII) has electronic configuration [Ar] or $1s^2\ 2s^22p^6\ 3s^23p^6$. This means when manganese is in oxidation state (VII) it has no d-electrons, so d–d transitions are not possible. Another transition metal ion which has no d-electrons and is coloured is the orange dichromate ion, $Cr_2O_7^{2-}$. Check for yourself that the orange colour of dichromate cannot be due to d–d transitions.

2 If you look at the example of visible spectra on the opposite page, you will see that both the complex ions contain Co^{2+} but the peaks of maximum absorbance are at different wavelengths. From the spectra suggest **(i)** the colours of these two complex ions and **(ii)** why the two ions have different peaks of maximum absorbance and different colours. (Hint: explain in terms of splitting of d orbitals.)

UNIT 1 PPAs 1–2

UNIT 1 PPA 1 – PREPARATION OF POTASSIUM TRIOXALATOFERRATE(III)

Aim

To prepare crystals of potassium trioxalatoferrate(III) and calculate the percentage yield.

Introduction

The trioxalatoferrate(III) complex ion has three oxalate ions bound to an iron(III) ion in an octahedral arrangement.

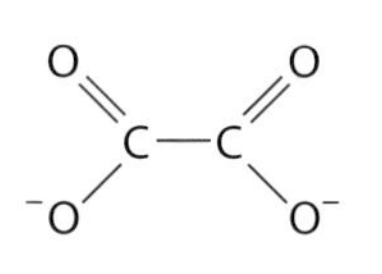

Oxalate ions are bidentate ligands (see page 24).

The starting reagent is hydrated ammonium iron(II) sulphate, which has formula, $(NH_4)_2Fe(SO_4)_2.6H_2O$. This contains the pale green Fe^{2+} ion. It is reacted with oxalic acid and, on warming, a light brown precipitate of iron(II) oxalate appears. The precipitate of iron(II) oxalate is filtered off and then treated with hydrogen peroxide and potassium oxalate solution, which are both colourless. The hydrogen peroxide oxidises iron(II) to iron(III) and some potassium trioxalatoferrate(III), together with a dark brown precipitate of iron(III) hydroxide, is produced.

More oxalic acid is then added and this converts the solid iron(III) hydroxide into more potassium trioxalatoferrate(III). The green solution of potassium trioxalatoferrate(III) is filtered and left aside in a dark cupboard. After a few days, lime green crystals of the product appear.

The chemical reactions are very complex but the overall stoichiometry is simply given by:
$(NH_4)_2Fe(SO_4)_2.6H_2O \rightarrow K_3[Fe(C_2O_4)_3].3H_2O$

This tells us that, in theory, one mole of the starting hydrated ammonium iron(II) sulphate should be converted to one mole of hydrated potassium trioxalatoferrate(III). This is important when calculating the % yield from your experimental results.

Results

The stoichiometry of the equation tells us that 1 mole of reactant gives 1 mole of product.

A sample set of results is given below:

Mass of hydrated ammonium iron(II) sulphate at start = 5·09 g
Mass of hydrated potassium trioxalatoferrate(III) formed = 2·65 g

The relevant equation is

$$(NH_4)_2Fe(SO_4)_2.6H_2O \rightarrow K_3[Fe(C_2O_4)_3].3H_2O$$

which shows that 1 mol $\leftrightarrow$ 1 mol

and so 392·1 g $\leftrightarrow$ 491·1 g

Theoretical yield calculation: $5{\cdot}09\,g \leftrightarrow \frac{491{\cdot}1 \times 5{\cdot}09}{392{\cdot}1} = 6{\cdot}38\,g$ if 100% conversion

$$\%\ yield = \frac{actual\ yield}{theoretical\ yield} \times 100 = \frac{2{\cdot}65}{6{\cdot}38} \times 100 = 41{\cdot}5\%$$

Conclusion

Green crystals of potassium hydrated trioxalatoferrate(III) were obtained and the yield was 41.5%.

Evaluation

DON'T FORGET

You need to be able to give reasons why the yield was not 100%.

Crystals of potassium trioxalatoferrate(III) were obtained so the procedure was effective.

Possible reasons for the yield being less than 100% include:

- the reactions are quite complex and some may have reached equilibrium rather than going to completion
- some side reactions may have taken place
- there may have been impurities in the initial reactants
- there may have been mass transfer losses
- there are always losses during crystallisation, since some of the product will remain in solution.

UNIT 1 PPA 2 – COLORIMETRIC DETERMINATION OF MANGANESE IN STEEL

Aim

To use colorimetry to find the percentage of manganese, by mass, in a steel paper clip.

Introduction

Colorimetry can be used to determine the concentrations of coloured substances in solution. A colorimeter consists essentially of a light source, a coloured filter, a light detector and recorder. The filter chosen is the complementary colour to the solution since this will result in maximum absorbance. The light passes through the filter then through the coloured solution. The difference in absorbance between the coloured solution and water is detected and noted as an absorbance value.

The first part of this PPA involves preparing and measuring the absorbance of different solutions of purple potassium permanganate at various concentrations. These solutions are prepared by accurately diluting a standard solution of 0·0010 mol l^{-1} potassium permanganate with deionised water. Measurements are made using a burette and standard flasks. The complementary colour to purple is green, so the colorimeter is fitted with a green filter. As the concentrations of the purple solutions decrease so do the absorbance values. From the results a calibration graph of absorbance against concentration can be drawn, similar to the one shown here.

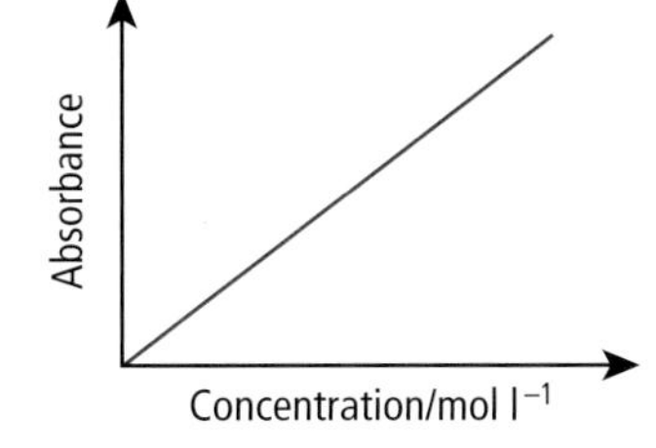

In the second part of the experiment, the manganese in the steel paper clip is converted into potassium permanganate solution. This is done by heating a known mass of a degreased paper clip with dilute nitric acid to get it into solution. The heating is carried out in a fume cupboard since brown fumes of the toxic gas, nitrogen dioxide, are evolved. The manganese in the steel is initially oxidised to Mn^{2+} ions by the nitric acid, then further oxidised to purple MnO_4^- ions by acidified potassium periodate. The resulting solution is transferred and made up to the mark in a 100 cm^3 standard flask, ensuring all the rinsings are also transferred. The solution should be light purple or pink. A sample of this solution is put into a cuvette and its absorbance measured using the colorimeter. The concentration corresponding to this absorbance can then be read from the calibration graph.

The ion-electron equations for the oxidation of manganese are:

$Mn \rightarrow Mn^{2+} + 2e^-$ and $Mn^{2+} + 4H_2O \rightarrow MnO_4^- + 8H^+ + 5e^-$

Results

A sample set of results is given below:

Initial mass of pieces of steel paper clip = 0·24 g
Absorbance of solution from the steel = 0·35
Concentration of this solution = $1{\cdot}32 \times 10^{-4}$ mol l^{-1} (read from the calibration graph)

The MnO_4^- solution from the steel had a total volume of 100 cm^3 and if its concentration was $1{\cdot}32 \times 10^{-4}$ moles in one litre then the number of moles of MnO_4^- in 100 cm^3 must have been $1{\cdot}32 \times 10^{-5}$.

This must also be the number of moles of Mn in the sample of the steel paper clip.

The mass of Mn in the sample = n × FM = $1{\cdot}32 \times 10^{-5}$ x 54·9 = 0·000725 g

Therefore the % of Mn in the steel = $\frac{0{\cdot}000725}{0{\cdot}24} \times 100 = 0{\cdot}30\%$

If you are given a sample set of results, you must be able to calculate the % Mn in the steel.

Conclusion

Colorimetry showed that the steel paper clip contained 0·30% manganese, by mass.

Evaluation

The main errors in this experiment were:

- incorrect measurements when diluting the standard potassium permanganate solution, thereby creating errors in the calibration graph values
- incorrect drawing of the best fit straight line in the calibration graph
- not all the manganese present in the steel clip may have been oxidised to MnO_4^- ions.

STOICHIOMETRY AND QUANTITATIVE REACTIONS

VOLUMETRIC ANALYSIS

A balanced chemical equation is a stoichiometric equation. It tells us the number of moles of reactants and the number of moles of products formed. A quantitative reaction is one in which the substances react completely according to the mole ratios given by the balanced equation.

Volumetric analysis involves using a solution of accurately known concentration in a quantitative reaction to determine the concentration of the other reactant. The procedure is known as **titration**. One solution is measured quantitatively into a conical flask using a pipette. The other solution is dispensed from a burette until a permanent colour change appears in the conical flask.

Normally a 'rough' titration is carried out, followed by more accurate titrations, until concordant results are obtained. Concordant results are within either 0·1 cm^3 or 0·2 cm^3 of each other, depending on the accuracy required. The mean or average value of the concordant results is used in calculations.

Standard solutions

A solution, the concentration of which is known accurately, is a **standard** solution. A standard solution can be prepared directly from a **primary standard**. This involves weighing out the primary standard accurately and dissolving it in deionised water in a beaker. The solution, plus all the rinsings, is then transferred into a standard flask and made up to the mark with more deionised water. A primary standard must have the following characteristics:

- be available in a high state of purity
- be stable when solid and when in solution
- be soluble in water
- have a reasonably high formula mass to reduce percentage errors when weighing.

DON'T FORGET

You must know the characteristics of a primary standard.

Acid–base titrations

The two types of titration that you have encountered so far are **acid–base** and **redox** titrations. During a titration, the experimenter looks for a permanent colour change in the conical flask, usually due to the presence of an indicator. This is known as the **end-point** in the reaction. The **equivalence** point is the point at which the reaction is *just* complete. The ideal situation is when the equivalence point and the end-point are exactly the same. Choosing the correct indicator and carrying out titrations very carefully and accurately help to ensure that the equivalence point and the end-point are the same. More information about acid–base indicators is given later in this book (see page 38).

DON'T FORGET

An indicator is a substance which changes colour at the end-point.

Redox titrations

During a redox titration an oxidising agent reacts with a reducing agent. The reducing agent is an electron donor and is itself oxidised during the reaction. The oxidising agent is an electron acceptor and is reduced. A very useful reagent in redox titrations is acidified permanganate (H^+/MnO_4^-) which is an excellent oxidising agent and has the advantage that it also acts as its own indicator.

Usually the purple potassium permanganate solution is placed in the burette and the reducing agent (plus some sulphuric acid which provides the H^+ ions) is in the conical flask. The purple permanganate ions change to colourless Mn^{2+} ions as they are added to the reducing agent. The end-point is observed when all the reducing agent has been used up and the purple MnO_4^- ions no longer react. The colour of the reaction mixture in the conical flask becomes a permanent light purple or pink colour.

contd

VOLUMETRIC ANALYSIS contd

Complexometric titrations

The third type of titration you have to be familiar with is **complexometric** titrations. These are based on the formation of complexes (see page 24). EDTA is a very important complexometric reagent and is used to determine the concentration of metal ions in solution (see Unit 2 PPA1, page 52).

GRAVIMETRIC ANALYSIS

Unlike volumetric analysis, the measurements made in gravimetric analysis are masses, normally determined using a digital balance accurate to two, three or four decimal places of a gram (g). In gravimetric analysis, the mass of an element or compound present in a substance is determined by changing that substance into another substance of known chemical composition and formula which can be readily isolated, purified and weighed. This may, for example, involve precipitation of silver(I) ions as silver chloride and, from the mass of silver(I) chloride precipitate filtered and dried, the original mass of silver(I) ions in the solution can be determined.

In gravimetric analysis, the final product usually has to be dried completely. This is done by the process of 'heating to constant mass' – heating the substance, allowing it to cool in a dry atmosphere inside a desiccator, then weighing it. The process is repeated until constant mass is achieved, showing that all the water has been driven off (see Unit 2 PPA2, page 54).

CALCULATIONS BASED ON VOLUMETRIC AND GRAVIMETRIC ANALYSIS

In most calculations associated with volumetric and gravimetric analysis, it is sensible to change the reacting quantities into moles. For substances in solution the formula used is:

$$n = V \times c$$

where n = number of moles, V = volume of solution in **litres** and c = concentration in mol l^{-1}.

For solids or pure substances where the mass has been measured the formula used is:

$$n = \frac{\text{mass}}{\text{FM}}$$

where n = number of moles, mass is measured in grams and FM is the formula mass of the substance.

DON'T FORGET

These two formulae for calculating the number of moles, n, are very important.

LET'S THINK ABOUT THIS

1 2·58 g of hydrated barium chloride, $BaCl_2.nH_2O$, was heated until constant mass was achieved, leaving 2·22 g of anhydrous barium chloride, $BaCl_2$. Calculate the value of n in the formula $BaCl_2.nH_2O$.

2 10·0 cm^3 of a liquid drain cleaner containing sodium hydroxide was diluted to 250 cm^3 in a standard flask. 25·0 cm^3 samples of this diluted solution were pipetted into a conical flask and titrated against 0·220 mol l^{-1} sulphuric acid solution. The average of the concordant titres was 17·8 cm^3. Calculate the mass of sodium hydroxide in 1 litre of the drain cleaner.

CHEMICAL EQUILIBRIUM

DYNAMIC EQUILIBRIUM

A chemical reaction is in dynamic equilibrium when the rate of the forward reaction is equal to the rate of the reverse reaction. At this stage, the concentrations of the reactants and the products will be constant but not equal.

From work done at Higher Level, you should know that the **position of equilibrium** can be changed by:

- altering the concentration of a reactant or product species
- changing the pressure if there are different numbers of moles of **gases** on both sides of the balanced equation
- altering the temperature.

You should also know that a catalyst speeds up the rate at which equilibrium is reached, but does not affect the position of equilibrium.

DON'T FORGET

It is important to appreciate that, in a dynamic equilibrium, both the forward reaction and the reverse reaction continue to take place, but at the same speed.

THE EQUILIBRIUM CONSTANT, K

The equilibrium constant is given the symbol, K, and for the general equation,

$aA + bB \rightleftharpoons cC + dD$ the equilibrium constant, $K = \frac{[C]^c[D]^d}{[A]^a[B]^b}$

where [A], [B], [C] and [D] are the equilibrium concentrations of A, B, C, and D respectively and a, b, c and d are the stoichiometric coefficients in a balanced reaction equation.

The balanced equation for the Haber process is

$N_2(g) + 3H_2(g) \rightleftharpoons 2NH_3(g)$

and the expression for the equilibrium constant is

$$K = \frac{[NH_3]^2}{[N_2][H_2]^3}$$

The concentration values are usually measured in $mol\,l^{-1}$, but for gaseous reactions partial pressures may be used. Whatever the concentrations are measured in, the value calculated for **K has no units**.

DON'T FORGET

You must be able to write the expression for the equilibrium constant from a balanced equation.

DON'T FORGET

Equilibrium constants have no units.

Homogeneous and heterogeneous equilibria

You are already familiar with the words 'homogeneous' and 'heterogeneous'. In a **homogeneous equilibrium** all the species are in the same phase. The equilibrium in the Haber process is an example of a homogeneous equilibrium.

In a **heterogeneous equilibrium** the species present are in more than one phase. An example of a heterogeneous equilibrium occurs when calcium carbonate is heated in a closed system. The carbon dioxide that forms cannot escape and an equilibrium is established. The equation is

$CaCO_3(s) \rightleftharpoons CaO(s) + CO_2(g)$

In a reaction like this – in which a pure solid or solids are present at equilibrium – the concentration of the solid is taken as being constant and is given the value of 1 in the equilibrium equation.

Therefore, in the above equilibrium, instead of $K = \frac{[CaO(s)][CO_2(g)]}{[CaCO_3(s)]}$
the correct expression is $K = [CO_2(g)]$

This is also true for pure liquids; their equilibrium concentration is also given the value 1 in the equilibrium expression. It is not true for aqueous solutions.

DON'T FORGET

Water is a pure liquid and so does not appear in the equilibrium constant expression; it is given the value 1.

contd

THE EQUILIBRIUM CONSTANT, K contd

Although, changing the concentration or pressure may affect the position of equilibrium, the equilibrium constant, K, does not vary at different concentrations or pressures. For example, If we consider the equilibrium in a solution of ammonia,

$NH_3(g) + H_2O(l) \rightleftharpoons NH_4^+(aq) + OH^-(aq)$

$$K = \frac{[NH_4^+][OH^-]}{[NH_3]}$$

If more ammonium ions in the form of solid ammonium chloride are added to this equilibrium mixture, the position of equilibrium shifts to the left since ammonium ions are present on the right-hand side of the equilibrium equation. This fits in with **Le Chatelier's principle** which states that '*When a reaction at equilibrium is subjected to a change the composition alters in such a way as to minimise the effects of that change.*'

This just means that, if the concentration of ammonium ions increases, the position of equilibrium shifts to decrease the concentration of the added ions, restoring equilibrium. However what, in effect, happens is that the position of equilibrium shifts until the ratio of the concentrations of the products to the concentrations of the reactants is the same as before, so that the value for K is re-established.

In the example above, when $[NH_4^+]$ increases some of the NH_4^+ ions react with OH^- ions to form more NH_3 and H_2O molecules until the previous value of K is restored.

EFFECT OF TEMPERATURE CHANGE ON THE VALUE OF K

Changes in concentration, pressure and the presence of a catalyst have no effect on the numerical value of K. However, the equilibrium constant **is temperature dependent**.

Consider the general equation, reactants $\rightleftharpoons$ products

For an endothermic reaction, an increase in temperature favours the products and so the ratio of $\frac{[products]}{[reactants]}$ increases and therefore K increases. A general statement is that for endothermic reactions, an increase in temperature causes an increase in the yield of the products and the value of K increases.

For an exothermic reaction, an increase in temperature favours the reactants and so the ratio of $\frac{[products]}{[reactants]}$ decreases and therefore K decreases. A general statement is that for exothermic reactions, an increase in temperature causes a decrease in the yield of the products and the value of K decreases.

You must know and understand the effect that changing the temperature has on the value of the equilibrium constant.

LET'S THINK ABOUT THIS

1 The equilibrium present in water is shown by the equation: $H_2O(l) \rightleftharpoons H^+(aq) + OH^-(aq)$
The value for the equilibrium constant, K, varies with temperature as shown in the table.

Temperature/°C	Equilibrium constant/K
0	$1{\cdot}14 \times 10^{-15}$
10	$2{\cdot}93 \times 10^{-15}$
25	$1{\cdot}01 \times 10^{-14}$
50	$5{\cdot}48 \times 10^{-14}$

(a) Write the expression for the equilibrium constant, K.
(b) The value of K increases as the temperature increases. What would be the sign of $\Delta H^{\ominus}$ for the forward reaction and, therefore, what type of reaction is it?

EQUILIBRIA BETWEEN DIFFERENT PHASES

PARTITION COEFFICIENT

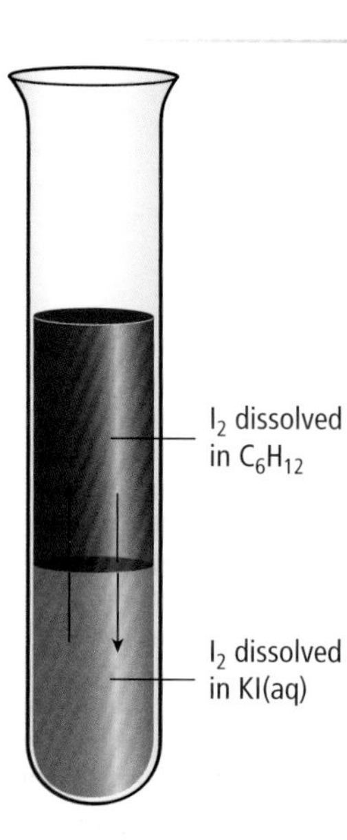

Solid iodine partitions between two immiscible liquids, cyclohexane and aqueous potassium iodide, until a dynamic equilibrium forms.

Immiscible liquids are liquids which do not mix together. The liquid with the lower density floats on the liquid with the greater density. For example, petrol floats on water, rather than mixing with it.

Another two immiscible liquids are cyclohexane and an aqueous solution of potassium iodide. Solid iodine is a solute which dissolves in both these immiscible liquids. When a solute such as iodine is shaken in two immiscible liquids, some of the solute dissolves in one solvent and the remainder dissolves in the other solvent. We say that the solute **partitions** itself between the two immiscible liquids.

Some of the solute that is dissolved in the lower layer starts to move into the upper layer. At the same time, some of the solute in the upper layer starts to move down into the lower layer. Eventually, the rate of upward movement of the solute is exactly the same as the rate of downward movement. At this stage, a dynamic equilibrium is established.

A particular solute will always partition itself between the same two solvents in the same ratio, as long as the temperature is the same. Since an equilibrium has been reached, it can be described by an equilibrium constant.

If the equation for the reaction is $I_2(aq) \rightleftharpoons I_2(C_6H_{12})$ then $K = \frac{[I_2(C_6H_{12})]}{[I_2(aq)]}$

This equilibrium constant is known as the **partition coefficient** and, as always with equilibrium constants, it is temperature dependent. However, is not affected by adding more solute or by increasing the volume of the solvents. However, it does depend on which solute and which solvents are used.

SOLVENT EXTRACTION

A separating funnel is used in solvent extraction.

Partition can be used to extract and purify a desired product from a reaction mixture. This technique is called solvent extraction and a separating funnel is used. The method depends on the desired material being more soluble in one liquid phase than another. For example, a cup of ordinary tea or coffee contains caffeine dissolved in water. However, caffeine is more soluble in an organic solvent such as dichloromethane, CH_2Cl_2.

Dichloromethane and water are immiscible solvents and so, if dichloromethane is added to a solution of caffeine in water, two layers form. Some of the caffeine in the water moves into the dichloromethane layer until equilibrium is reached. Since there are two layers, the mixture can be separated using a separating funnel; the lower layer is first run off into one container and then the upper layer is run off into another container.

Decaffeination of coffee is achieved by solvent extraction. Originally, dichloromethane was used as the solvent but a more environmentally friendly solvent called supercritical carbon dioxide is now used. Supercritical carbon dioxide exists in closed containers above 73 atmospheres and 31°C. It has the advantage that it behaves both like a gas and a liquid. When passed through green coffee beans, the gas penetrates the beans and its liquid properties allow it to dissolve out about 98% of the caffeine.

Solvent extraction is also used to purify water-soluble organic acids using a suitable organic solvent.

It can be shown by calculation that it is more efficient to use smaller quantities of solvent a few times, rather than the same total volume once only. (See the 'Let's think about this' section.)

CHROMATOGRAPHY

Using chromatography to separate substances in a mixture also depends on the partition equilibrium between two phases, one stationary and the other mobile. In paper chromatography the stationary phase is water present on the very absorbent chromatography paper, not the paper itself. The mobile phase is the solvent used. The mobile phase flows through the stationary phase and carries the components of the mixture with it.

Different components in the mixture travel at different rates and are thus separated. Some compounds in a mixture travel almost as far as the solvent does; some hardly move at all. The distance travelled relative to the solvent is constant for a particular compound, as long as the same type of paper and the same solvent mixture are used.

The distance travelled relative to the solvent is called the R_f value. For each compound it can be worked out using the expression:

$$R_f = \frac{\text{distance travelled by compound}}{\text{distance travelled by solvent}}$$

Thin layer chromatography is often used as a faster alternative to paper chromatography. Instead of paper, a thin uniform layer of silica gel or alumina coated onto a piece of glass or plastic is used. Water held on the silica gel (or the alumina) is the stationary phase. The mobile phase is the liquid solvent or mixture of solvents.

Gas–liquid chromatography (often called simply 'gas chromatography') is a form of chromatography encountered less often in schools and colleges, since gas–liquid chromatography machines are expensive. Here, the mobile phase is an unreactive gas, such as helium, which is known as the carrier gas. The stationary phase is a liquid held onto a solid and contained in a column coiled up inside an oven. The sample is injected into the column and the oven temperature is set so that the substances within the sample vapourise. They are carried through the column as gases. The speed that a particular compound travels through the column depends on how much time is spent moving with the mobile carrier gas, as opposed to being attached to the stationary liquid.

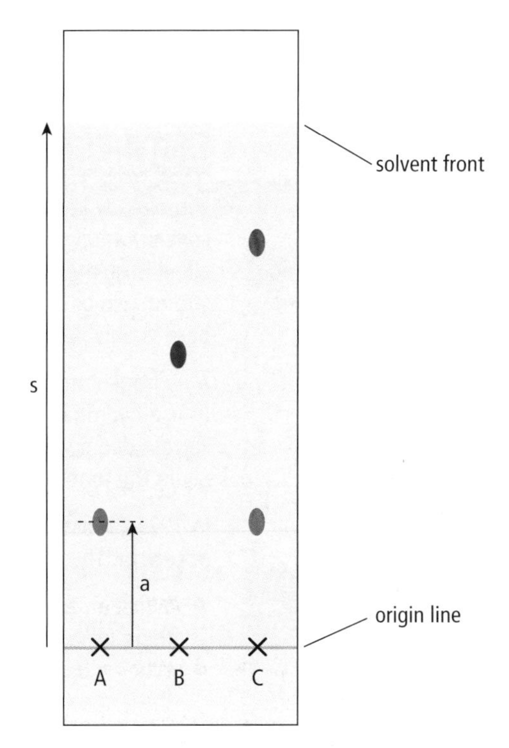

This diagram shows a completed paper chromatogram. At the start A, B and C substances had been spotted onto the origin line at three locations shown by letter X's. The solvent never travels up the paper in a straight line and it can be difficult to measure the exact distance travelled by the solvent front.
The R_f value for substance A is given by $R_f = \frac{a}{s}$
It can be seen from this diagram that spot C contains two components, one of which is substance A. The other component of C is not substance B since the R_f values will be different.

You will find more information about these different forms of chromatography at http://www.chemguide.co.uk/analysis/chromatography/paper.html http://www.chemguide.co.uk/analysis/chromatography/thinlayer.html http://www.chemguide.co.uk/analysis/chromatography/gas.html

LET'S THINK ABOUT THIS

For the equilibrium, caffeine(aq) $\rightleftharpoons$ caffeine(CH_2Cl_2) $K = \frac{[\text{caffeine}(CH_2CL_2)]}{[\text{caffeine(aq)}]} = 4{\cdot}6$

(This means that caffeine is 4·6 times more soluble in CH_2Cl_2 than in water.)

100 cm^3 of a solution of tea contains 0·30 g of caffeine and 100 cm^3 of dichloromethane is available for extraction. Calculate the difference in the quantity of caffeine extracted in two consecutive extractions using 50 cm^3 of dichloromethane each time, compared to using 100 cm^3 of dichloromethane in one extraction.

EQUILIBRIA INVOLVING IONS

EQUILIBRIA IN ACIDS

Very often hydrogen ions are simply written as H^+ but you must appreciate that they are always hydrated.

You already know that the concentration of hydrogen ions in an acid is greater than the concentration of hydrogen ions in water. Likewise, the concentration of hydroxide ions in an alkali is greater than the concentration of hydroxide ions in water. You are also aware that a strong acid or strong base is fully ionised in aqueous solution and that a weak acid or weak base is only partially ionised in aqueous solution.

A hydrogen atom contains a proton and an electron; so a hydrogen ion is really just a proton (it is a hydrogen atom that has lost an electron). Hydrogen ions, as such, only exist when surrounded by water molecules in an aqueous solution. These hydrated hydrogen ions are often given the formula $H^+(aq)$ or H_3O^+ or even $H_9O_4^+$.

You need to be able to identify conjugate acids and conjugate bases from equations showing the dissociation of acids in water.

A useful website is http://dl.clackamas.edu/ch105-04/conjugat.htm

In 1923 Bronsted and Lowry defined an **acid as a proton donor** and a **base as a proton acceptor**. These are known as the Bronsted–Lowry definitions of an acid and base.

- When an acid donates a proton, the species left is called the **conjugate base**. For every acid there is a conjugate base that is formed when the acid loses a proton.
- When a base accepts a proton, the species left is called the **conjugate acid**. For every base there is a conjugate acid that is formed when the base gains a proton.

The table gives examples.

acid +	base →	conjugate base +	conjugate acid
HNO_3	H_2O	NO_3^-	H_3O^+
HCOOH	H_2O	$HCOO^-$	H_3O^+
H_2O	NH_3	OH^-	NH_4^+
H_2O	F^-	OH^-	HF

IONISATION OF WATER

As you can see in the table above, water acts both as an acid and a base; water is amphoteric. It can donate a proton leaving OH^- and can accept a proton to form H_3O^+.

The ionisation of water can be thought of as one water molecule acting as an acid reacting with another water molecule acting as a base to form the conjugate base and conjugate acid. This is also known as the dissociation of water and is represented by the equation:

Since water is a pure liquid, $[H_2O]$ is given the value 1.

$$\underset{\text{acid}}{H_2O(l)} + \underset{\text{base}}{H_2O(l)} \rightleftharpoons \underset{\text{conjugate acid}}{H_3O^+(aq)} + \underset{\text{conjugate base}}{OH^-(aq)}$$

The expression for the equilibrium constant is $K = [H_3O^+(aq)] \times [OH^-(aq)]$
or more simply $K = [H^+][OH^-]$

Since K_w is an equilibrium constant, it has no units.

This equilibrium constant or dissociation constant is known as the **ionic product** of water and is given the symbol, K_w.

Since K_w is an equilibrium constant, its value is temperature dependent. At 25°C the value of K_w is approximately 1×10^{-14}. You can see different values for K_w at different temperatures on page 33.

THE pH SCALE

- Acidic solutions have pH values lower than 7 and $[H^+] > [OH^-]$.
- Alkaline or basic solutions have pH values above 7 and $[H^+] < [OH^-]$.
- Neutral solutions have pH = 7 and in a neutral solution, $[H^+] = [OH^-]$.

The greater the concentration of hydrogen ions, the lower is the pH value. In fact, the relationship between pH and the hydrogen ion concentration is given by $pH = -\log[H^+]$

contd

THE pH SCALE contd

The pH of neutral solutions can be calculated to be 7 as shown below.

$K_w = [H^+] \times [OH^-] = 1 \times 10^{-14}$ and in pure water $[H^+] = [OH^-]$ and so $K_w = [H^+]^2$
Therefore $[H^+] = \sqrt{1 \times 10^{-14}} = 1 \times 10^{-7}\,mol\,l^{-1}$
$pH = -\log[H^+] = -\log(10^{-7}) = 7$
pH values are not always whole numbers. They can be non-integral values.

Worked examples

Calculate the pH of 0·21 mol l^{-1} HCl(aq).

HCl is a strong acid and, since it is fully ionised in aqueous solution, the hydrogen ion concentration is the same as the original concentration of the acid. So, $[H^+] = 0{\cdot}21\,mol\,l^{-1}$
$pH = -\log(0{\cdot}21) = -(-0{\cdot}68) = 0{\cdot}68$

You should also be able to calculate the hydrogen ion concentration from a non-integral pH value. For example, calculate the hydrogen ion concentration in a solution of pH = 11·9.

$pH = -\log[H^+] = 11{\cdot}90$
So, $\log[H^+] = -11{\cdot}90$ and $[H^+] = 1{\cdot}3 \times 10^{-12}\,mol\,l^{-1}$

DON'T FORGET

You must be able to do these calculations. Make sure you know how to use your calculator so that you get the correct answer every time.

pH AND WEAK ACIDS

For a strong **monoprotic** acid such as HCl or HNO_3 the hydrogen ion concentration, $[H^+]$, is the same as the original concentration of the acid. This is because a strong acid is fully ionised, so all the acid molecules have changed into ions. So in 0·25 mol l^{-1} HCl(aq), $[H^+] = 0{\cdot}25\,mol\,l^{-1}$.

However, weak acids are only partially ionised in aqueous solution. In fact, for most weak acids less than 1% of the acid molecules dissociate into ions in aqueous solution. This means that, for weak acids, the hydrogen ion concentration in solution will be much less than the concentration of the acid. So how can we calculate the pH and $[H^+]$ for a weak acid in solution?

If we consider any weak acid and give it the general formula HA, the dissociation of the weak acid can be represented by:

$$\underset{\text{acid}}{HA(aq)} + \underset{\text{base}}{H_2O(l)} \rightleftharpoons \underset{\text{conjugate acid}}{H_3O^+(aq)} + \underset{\text{conjugate base}}{A^-(aq)}$$

This can be written more simply as $HA \rightleftharpoons H^+ + A^-$ and the equilibrium constant which is known as the dissociation constant of the acid, symbol K_a is given by $K_a = \frac{[H^+][A^-]}{[HA]}$

For a weak acid, $[H^+] = \sqrt{K_a c}$ where c is the concentration of the acid.

This allows you to calculate K_a from $[H^+]$, and $[H^+]$ from K_a if the concentration of the weak acid is known. If $[H^+]$ is calculated, the pH can be worked out using $pH = -\log[H^+]$

Just as $pH = -\log[H^+]$, the dissociation constant of an acid can be represented by pK_a where
$\mathbf{pK_a = -\log K_a}$

It is often more convenient to use pK_a instead of K_a. A very useful expression which shows the relationship of the pH of a weak acid to its dissociation constant is $\mathbf{pH = \frac{1}{2}pK_a - \frac{1}{2}\log c}$ where c is the concentration of the acid.

DON'T FORGET

Values of acid dissociation constants, K_a and pK_a, are given on page 12 of the SQA Data Booklet.

DON'T FORGET

You must know these formulae and be able to use them in calculations.

LET'S THINK ABOUT THIS

1. Write down the conjugate acid and conjugate base for ethanoic acid.
2. Calculate the pH of **(a)** 0·22 mol l^{-1} nitric acid and **(b)** 0·12 mol l^{-1} sodium hydroxide.
3. Calculate the hydrogen ion concentration when the pH is **(a)** 4·3 and **(b)** 8·3.
4. Calculate the pH of **(a)** 0·1 mol l^{-1} ethanoic acid and **(b)** 0·2 mol l^{-1} benzoic acid. (Remember that pK_a and K_a values are given in the SQA Data Booklet page 12.)

BASES, INDICATORS AND BUFFER SOLUTIONS

DON'T FORGET

A base is a proton acceptor and a hydrogen ion is a proton.

DON'T FORGET

The solvent water is a pure liquid and so does not appear in the expression for the equilibrium constant.

EQUILIBRIA IN BASES

Like weak acids, weak bases are only partially ionised, or dissociated, in aqueous solution. The equation which represents a base of general formula B dissociating in aqueous solution is:

$B(aq) + H_2O(l) \rightleftharpoons BH^+(aq) + OH^-(aq)$

An example of a weak base is ammonia, NH_3.

$NH_3(aq) + H_2O(l) \rightleftharpoons NH_4^+(aq) + OH^-(aq)$

Here, NH_3 is the base, H_2O is the acid and the conjugate acid is the ammonium ion, NH_4^+. The conjugate base is OH^-.

The dissociation constant (K_b) for ammonia $= \dfrac{[NH_4^+][OH^-]}{[NH_3]}$

The ammonium ion formed is a weak acid and will dissociate as in the equation:

$NH_4^+(aq) + H_2O(l) \rightleftharpoons NH_3(aq) + H_3O^+(aq)$

This time, H_2O is the base, the conjugate acid is H_3O^+ and the conjugate base is NH_3.

The dissociation constant for the ammonium ion, $K_a = \dfrac{[NH_3][H_3O^+]}{[NH_4]^+}$

The greater the numerical value of K_a for a weak acid, the stronger it is. The value of K_a for the ammonium ion is $5{\cdot}8 \times 10^{-10}$, so it is much weaker than ethanoic acid which has $K_a = 1{\cdot}7 \times 10^{-5}$.

- The greater the value of K_a for a species, the more acidic and the less basic it is.
- The lower the value of K_a ,the less acidic it is and the more basic it is.

INDICATORS

Indicators are used in acid–base titrations since they change colour at the end-point of the reaction. Indicators are usually weak acids in which the colour of the acid is different from its conjugate base. For example, if we represent the indicator as a weak acid of formula HIn (red in colour), and its conjugate base as In^- (blue), its dissociation in water can be represented as:

$HIn(aq) + H_2O(l) \rightleftharpoons H_3O^+(aq) + In^-(aq)$
red — blue

- Adding an acid (i.e. adding H_3O^+ ions) will shift the position of equilibrium to the left and the solution will become red in colour.
- Adding an alkali (i.e. adding OH^- ions) will remove hydrogen ions from the above equilibrium and the position of equilibrium will shift to the right; the colour will now become blue.

For a substance to be a useful indicator, the colour of the un-ionised form, HIn, must be distinctly different from that of its conjugate base, In^-.

The expression for the equilibrium constant for the equilibrium above is:

$K_{In} = \dfrac{[H_3O^+(aq)][In^-(aq)]}{[HIn(aq)]}$ which can be rearranged to $\dfrac{[In^-]}{[HIn]} = \dfrac{K_{In}}{[H_3O^+]}$

The colour of the indicator at any time during a titration depends on the relative concentrations of the un-ionised indicator molecules and its conjugate base, that is $\dfrac{[In^-]}{[HIn]}$

This will be equal to 1 when $\dfrac{K_{In}}{[H_3O^+]} = 1$, i.e. when $pK_{In} = pH$

This tells us that the pK_{In} (the pK of the indicator) is equal to the pH when the indicator changes colour. The colour change is only distinguishable when the [HIn] and [In^-] differ by a factor of approximately 10 (so, only when the [HIn] is approximately 10 times greater than [In^-] or vice-versa). This means the pH range over which the colour change of the indicator occurs is approximately $pH = pK_{In} \pm 1$; the approximate range of most indicators is over two pH units.

contd

INDICATORS contd

The pH range of an indicator chosen for a titration must coincide with the point at which the pH is changing very rapidly. The pH changes during a titration occur most rapidly around the end-point. This is seen in three of the four diagrams below.

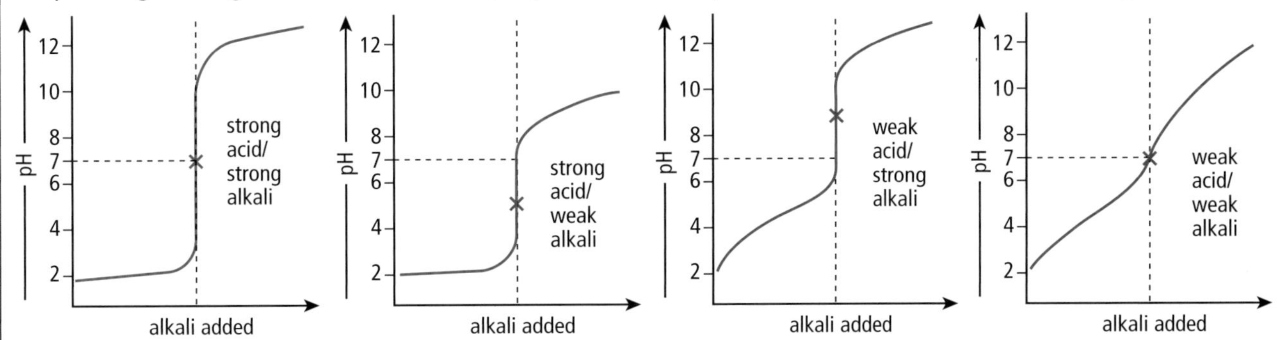

To be useful in a titration, the pH range of an indicator must lie within the 'vertical' part of the graph. For a strong acid/strong alkali titration, the vertical part lies between approximately pH 3 to pH 11 – the pH range of most indicators fit into this range. For a strong acid/weak alkali titration the pH range of the indicator should be between approximately pH 3 and 7. For a weak acid/strong alkali titration, the pH range of the indicator should lie between approximately pH 7 and 11. The vertical part of a weak acid/weak alkali titration is almost non-existent so no indicator is suitable for a weak acid/weak alkali titration.

BUFFERS

A buffer solution is one in which the pH remains approximately constant when small amounts of acid or base or water are added.

An acid buffer consists of a solution of a weak acid and one of its salts. An example of an acid buffer is a solution of ethanoic acid and sodium ethanoate. Sodium ethanoate is an ionic salt and, although the sodium ions are just spectator ions, the ethanoate ions are the conjugate base of ethanoic acid. Ethanoic acid is a weak acid and is only partially ionised:

$CH_3COOH(aq) \rightleftharpoons CH_3COO^-(aq) + H^+(aq)$

When **an acid is added** to this buffer solution the added H^+ ions from the acid react with the CH_3COO^- ions **from the salt** forming more un-ionised CH_3COOH acid molecules. The overall result is that the concentration of H^+ ions stays the same, so the pH remains unchanged.

When **an alkali is added** to this buffer the added OH^- ions react with the hydrogen ions from the ethanoic acid. More ethanoic acid molecules dissociate to replace these hydrogen ions. Again, the overall result is that the concentration of H^+ ions stays the same, so the pH remains unchanged.

A basic buffer consists of a solution of a weak base and one of its salts, such as a solution of ammonia and ammonium chloride. The weak base, ammonia, removes any added hydrogen ions. The conjugate acid, the ammonium ions from the ammonium chloride salt, replace any hydrogen ions removed when an alkali is added.

The pH of an acid buffer solution can be calculated using the expression: $\mathbf{pH = pK_a - \log \frac{[acid]}{[salt]}}$

Using this expression, the required composition of an acid buffer solution can be calculated for a particular pH.

The table found at the website below gives the pH ranges and colours of a larger number of different indicators. Note the American spelling, though. http://chemistry.about.com/library/weekly/aa112201a.htm

DON'T FORGET

You need to know the definition of a buffer and how it works. Also you must know and be able to do calculations using this expression.

LET'S THINK ABOUT THIS

1 Explain which indicators from the table would be most useful for **(a)** a weak acid/strong alkali titration and **(b)** a strong acid/weak alkali titration.

2 What could be added to a solution of propanoic acid to make an acid buffer?

3 Explain how a solution of methanoic acid and sodium methanoate acts as a buffer when small quantities of the following are added: **(a)** an acid **(b)** an alkali **(c)** water.

4 Use the expression above and information from the SQA Data Booklet to calculate the pH of the buffer solution formed by mixing 40 cm^3 of 0·1 mol l^{-1} ethanoic acid and 60 cm^3 of 0·1 mol l^{-1} sodium ethanoate solution.

Indicator	pH range
methyl red	4·4–6·2
phenolphthalein	8·0–10·0
bromocresol green	4·0–5·6
methyl orange	3·1–4·4

THERMOCHEMISTRY, HESS'S LAW AND BOND ENTHALPIES

You need to know these definitions and be able to write the balanced equation corresponding to the enthalpy of formation of different substances. Remember:
$\Delta H^\ominus$ = –ve for an exothermic reaction
$\Delta H^\ominus$ = +ve for an endothermic reaction

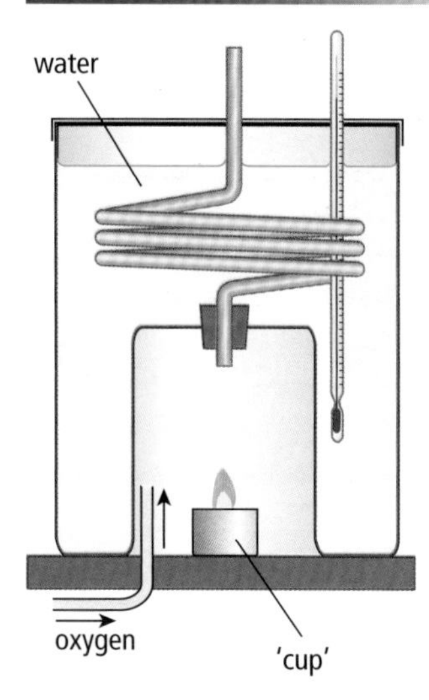

This is a more sophisticated calorimeter than the copper-can arrangement you might have used.

The heat given out by the burning material, E_h, can be calculated using $cm\Delta T$.

ENTHALPY CHANGES

Most chemical reactions, such as neutralisation and combustion, are **exothermic**; some, such as sparking air to produce oxides of nitrogen, are **endothermic**. Thermochemistry concerns the study of energy changes which take place during chemical reactions. Every substance has an enthalpy value but this cannot be measured. Only enthalpy **changes** can be measured.

The 'standard' enthalpy change, $\Delta H^\ominus$, refers to the enthalpy change for a reaction in which the reactants and products are in their **standard states** at a specified temperature. The standard state is the most stable state of the substance under **standard conditions**: one atmosphere pressure and a specified temperature, usually 25°C (298K).

Definitions of standard enthalpies that you need to know are:

- The **standard enthalpy of combustion of a substance, $\Delta H^\ominus_c$**, is the energy given out when **one mole** of the substance burns completely in oxygen. The equation which refers to the standard enthalpy of combustion of ethanol is

$$C_2H_5OH(l) + 3O_2(g) \rightarrow 2CO_2(g) + 3H_2O(l)$$

- The **standard enthalpy of formation of a compound, $\Delta H^\ominus_f$**, is the energy given out or taken in when **one mole** of the compound is formed from its **elements in their standard states**. The equation which refers to the standard enthalpy of formation of ethanol is

$$2C(s) + 3H_2(g) + \tfrac{1}{2}O_2(g) \rightarrow C_2H_5OH(l)$$

Standard enthalpy changes, $\Delta H^\ominus$, are usually measured in units of $kJ\,mol^{-1}$ of reactant or product and the enthalpy change for a reaction is $\Delta H^\ominus = \Sigma\Delta H^\ominus_{products} - \Sigma\Delta H^\ominus_{reactants}$

- For endothermic reactions, $\Delta H^\ominus$ is positive and the enthalpy of the products is greater than that of the reactants.
- For exothermic reactions, $\Delta H^\ominus$ is negative and the enthalpy of the products is less than that of the reactants.

Calorimetry

You will remember from Higher Chemistry the PPA experiment on enthalpy of combustion. In the experiment, ethanol was burned in a spirit burner below a copper can containing water. The copper can is known as a calorimeter and the quantitative determination of the change in heat energy which occurs during a chemical reaction is known as **calorimetry**. So, a calorimeter is used to measure the quantity of heat energy given out or taken in during a chemical reaction.

A small quantity of combustible material is placed inside the calorimeter in a small 'cup' surrounded by oxygen gas. The sample is ignited electrically. The heat that is produced passes through the copper coil into the water and the temperature rise of the water is measured using an accurate thermometer.

HESS'S LAW

Hess's Law states that the enthalpy change for a particular reaction depends only on the enthalpies of the starting reactants and of the final products, but is independent of the route taken. The overall reaction enthalpy is the sum of the reaction enthalpies of each step in the reaction. This is a direct consequence of the First Law of Thermodynamics which states that, in any chemical reaction, energy must be conserved. This means that energy can neither be created nor destroyed in a chemical reaction.

In calculations (assuming we know the value, in $kJ\,mol^{-1}$, for the enthalpy change corresponding to the equation for the chemical reaction) then

- if the equation is multiplied, the enthalpy value has to be multiplied by the same amount
- if the equation is reversed , the sign of the enthalpy change must change.

Some calculations using Hess's Law are given in the 'Let's think about this' section.

BOND ENTHALPIES

Energy is required to break a bond. Using Hess's Law, the same amount of energy must be given out when that bond is made; bond-breaking is endothermic and bond-making is exothermic.

The table on the right-hand side of page 9 in the SQA Data Booklet is for diatomic molecules. For example, the bond enthalpy of H–F is 569 kJ mol^{-1}. This means that 569 kJ is required to break $6{\cdot}02 \times 10^{23}$ (1 mol) H–F bonds and 569 kJ is given out when $6{\cdot}02 \times 10^{23}$ (1 mol) H–F bonds are formed.

The table on the left-hand side of page 9 in the SQA Data Booklet shows the mean or average bond enthalpies. These are for bonds which occur in different environments. The C–H bond can occur in alkanes, alkenes and aldehydes, for example. The C–H bonds in these different types of molecule have slightly different bond enthalpies.

The mean bond enthalpy for the O–H bond is 458 kJ mol^{-1}.
Therefore, $\Delta H^{\ominus}$ for the reaction $H_2O(g) \rightarrow O(g) + 2H(g)$ is $2 \times 458 = 916$ kJ mol^{-1} since two moles of O–H bonds are being broken for every mole of $H_2O(g)$.

When doing calculations based on bond enthalpy data, it is important that we consider the reactants and products in the gas state. This means that there are no other interactions involved, just the covalent bonds which are breaking or being made. However, carbon is a solid at room temperature. So, if carbon solid is present as a reactant, we must also take into account the change in enthalpy required to change the solid carbon into the gas state.

Conversion of a solid directly into the gas state is known as sublimation and the **enthalpy of sublimation of carbon** is also given on page 9 of the SQA Data Booklet. The value quoted of 715 kJ mol^{-1} is for $C(s) \rightarrow C(g)$, i.e. for the conversion of one mole of solid carbon with a covalent network structure into one mole of individual carbon atoms in the gas state.

Worked example

Calculate the enthalpy of formation of ethene using bond enthalpy data from the SQA Data Booklet, page 9.

The target equation is $2C(s) + 2H_2(g) \rightarrow C_2H_4(g)$
Using structural formulae this becomes $2C(s) + 2(H{-}H)(g) \rightarrow H_2C{=}CH_2(g)$
From this, you can see that the bond-breaking and bond-making steps are:

Bond-breaking ($\Delta H^{\ominus}$ +ve)
$2C(s) \rightarrow 2C(g)$ $\Delta H^{\ominus} = 2 \times 715 = 1430$ kJ
$2(H{-}H)(g) \rightarrow 4H(g)$ $\Delta H^{\ominus} = 2 \times 432 = 864$ kJ

Bond-making ($\Delta H^{\ominus}$ –ve)
$1 \times$ C=C $\Delta H^{\ominus} = -602$ kJ
$4 \times$ C–H $\Delta H^{\ominus} = 4 \times -414 = -1656$ kJ

Therefore, the overall enthalpy change is
$1430 + 864 - 602 - 1656 = +36$ kJ mol^{-1}

The value given in the SQA Data Booklet for the enthalpy of formation of ethene is + 52 kJ mol^{-1}.

Since some of the bond enthalpy values given in the SQA Data Booklet are mean values, the enthalpies of reactions calculated from bond enthalpy values often differ from those determined experimentally, including those calculated from enthalpy of combustion data.

DON'T FORGET

Bond enthalpy or bond dissociation values are given on two different tables on page 9 of the SQA Data Booklet.

DON'T FORGET

$H{-}F(g) \rightarrow H(g) + F(g)$
$\Delta H^{\ominus} = 569$ kJ mol^{-1}
$H(g) + F(g) \rightarrow H - F(g)$
$\Delta H^{\ominus} = -569$ kJ mol^{-1}

DON'T FORGET

When doing calculations using bond enthalpy data, you must consider all the bonds to be broken and all the bonds to be made. It is best to draw out the structural formulae of the reactants and products so that you see all the bonds involved.

DON'T FORGET

Drawing out the full structural formulae helps you see all the bonds which are to be broken and made.

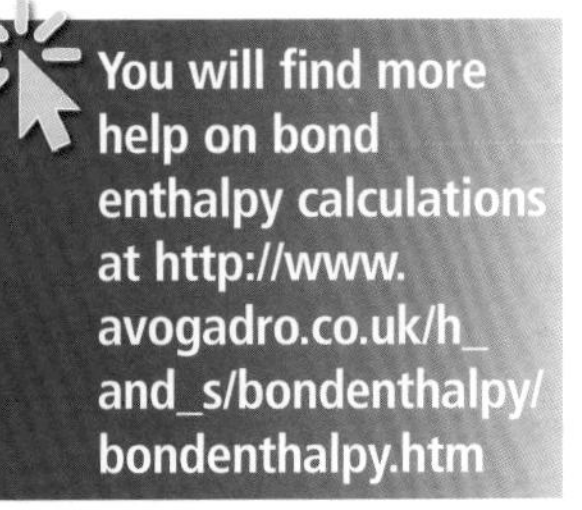

LET'S THINK ABOUT THIS

1. Using enthalpy of combustion data from page 9 of the SQA Data Booklet, calculate the enthalpy of formation of **(a)** ethanol and **(b)** ethyne.
2. Using bond enthalpy data from page 9 of the SQA Data Booklet calculate the enthalpy of formation of **(a)** ethanol and **(b)** ethyne.
3. Suggest why the values calculated in questions 1 and 2 are very different.

HESS'S LAW APPLIED TO IONIC SUBSTANCES

BORN–HABER CYCLE

The Born–Haber cycle is a thermochemical cycle which is applied to the formation of an ionic compound. For example, let us examine all the *hypothetical* steps involved in the formation of calcium iodide from its elements in their standard states: $Ca(s) + I_2(s) \rightarrow Ca^{2+}(I^-)_2(s)$

Taking calcium first, we have to consider all the steps involved in converting one mole of solid calcium metal into one mole of individual gaseous Ca^{2+} ions. The calcium metallic lattice has to be broken to form individual atoms of calcium. These individual calcium atoms have to be in the gas state, so that there are no interactions between them.

The energy required to do this is the standard molar **enthalpy of atomisation** and is **the energy required to produce one mole of isolated gaseous atoms** from the element in its standard state. For calcium, the equation which shows this is $Ca(s) \rightarrow Ca(g)$ and the enthalpy value from the table on page 17 of the SQA Data Booklet is $+178\,kJ\,mol^{-1}$.

The Born–Haber cycle below shows the values given in the text on the right for the different steps involved (note that the diagram is not to scale).

Next, the calcium atoms have to be converted into ions. The **first ionisation energy** of calcium is the energy required to convert one mole of calcium atoms in the gas state into one mole of Ca^+ ions in the gas state. The equation which shows this is $Ca(g) \rightarrow Ca^+(g) + e^-$ and the value given for this in the SQA Data Booklet is $+596\,kJ\,mol^{-1}$.

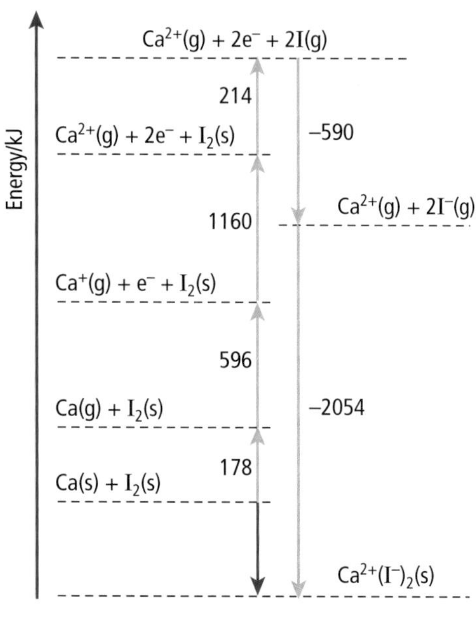

Using this Born–Haber cycle, it can be calculated that the overall $\Delta H^\ominus$ for the formation of calcium iodide is $-496\,kJ\,mol^{-1}$.

The calcium ions in calcium iodide are Ca^{2+} ions, so the next step is to consider the **second ionisation energy** of calcium which is the energy required to convert one mole of Ca^+ ions in the gas state into one mole of Ca^{2+} ions in the gas state. The equation which shows this is $Ca^+(g) \rightarrow Ca^{2+}(g) + e^-$ and the value given for this in the SQA Data Booklet is $+1160\,kJ\,mol^{-1}$.

Next we have to consider the enthalpy changes which take place when one mole of solid iodine changes into two moles of iodide ions. The first step is to find the standard molar enthalpy of atomisation of solid iodine. The equation for this is $\frac{1}{2}I_2(s) \rightarrow I(g)$ and the value given for this in the SQA Data Booklet is $107\,kJ\,mol^{-1}$.

However, since one mole of solid iodine is present as a reactant, the value of $\Delta H^\ominus$ for the first step is $+214\,kJ$. The individual iodine atoms then have to gain one electron each to form negative iodide ions, still in the gas state, $I^-(g)$. This energy change is known as the electron affinity. The **electron affinity** is defined as the enthalpy change when one mole of electrons is added to one mole of isolated atoms in the gas state. For iodine, the value of the electron affinity is $-295\,kJ\,mol^{-1}$, referring to the reaction $I(g) + e^- \rightarrow I^-(g)$

Therefore, for two moles of $I(g)$ to change into two moles of $I^-(g)$ the enthalpy change will be $-590\,kJ$.

The table at the top of page 18 in the SQA Data Booklet gives values of electron affinities for some of the more common non-metallic elements.

In the formation of one mole of calcium iodide from its elements in their standard states the sum of the enthalpy changes we have discussed so far is
$\Delta H^\ominus = 178 + 596 + 1160 + 214 - 590 = +1558\,kJ$

At this stage, the process looks to be endothermic. What makes the overall process exothermic is the large negative value for the lattice enthalpy. This value $-2054\,kJ\,mol^{-1}$ is not given on page 17 of the Data Booklet but the values for many other ionic compounds are. The equation which represents the lattice enthalpy of calcium iodide is: $Ca^{2+}(g) + 2I^-(g) \rightarrow Ca^{2+}(I^-)_2(s)$

Most Born–Haber cycles you will come across include the bond enthalpy value for the non-metal; for example, the Cl–Cl bond would have to broken in the formation of Na^+Cl^- or $Mg^{2+}(Cl^-)_2$ and the bond enthalpy value for the Cl–Cl bond would have to be used.

You need to know the names of the different steps in a Born–Haber cycle and be able to find their values in the SQA Data Booklet.

You will find some more Born–Haber cycles at http://www.tutorvista.com/content/chemistry/chemistry-iii/chemical-bonding/born-haber-cycle.php but note that the enthalpy values may be different from those quoted in the SQA Data Booklet.

ENTHALPY OF SOLUTION

Just as the Born–Haber cycle is a thermochemical cycle applied to the formation of an ionic crystal, a similar thermochemical cycle can be used to determine the enthalpy of solution when one mole of an ionic crystal dissolves in water.

Consider the steps involved when an ionic crystal, such as magnesium chloride, dissolves in water.

The equation which represents the enthalpy of solution of magnesium chloride is
$Mg^{2+}(Cl^-)_2(s) \rightarrow Mg^{2+}(aq) + 2Cl^-(aq)$

Firstly, the ionic lattice has to be broken:
$Mg^{2+}(Cl^-)_2(s) \rightarrow Mg^{2+}(g) + 2Cl^-(g)$

This is the reverse of the lattice formation process. Looking at the SQA Data Booklet, page 17, the lattice enthalpy of magnesium chloride is given as $-2326\,kJ\,mol^{-1}$. Therefore, for the reverse process, $\Delta H^{\ominus}$ will be $+2326\,kJ\,mol^{-1}$.

The next step is converting the ions in the gas state into hydrated ions. This is known as hydration. The hydration enthalpy is the energy released when one mole of individual gaseous ions becomes hydrated. Values for hydration enthalpies of some positive and negative ions are given on page 18 of the SQA Data Booklet. The value for the hydration enthalpy of magnesium ions is $-1920\,kJ\,mol^{-1}$ and the equation which refers to this reaction is $Mg^{2+}(g) \rightarrow Mg^{2+}(aq)$

The value for the hydration enthalpy of chloride ions is $-364\,kJ\,mol^{-1}$ and the equation which refers to this reaction is $Cl^-(g) \rightarrow Cl^-(aq)$

However, one mole of magnesium chloride contains two moles of chloride ions, so the reaction is $2Cl^-(g) \rightarrow 2Cl^-(aq)$ and the enthalpy change is $-728\,kJ$.

The overall calculation for the enthalpy of solution of magnesium chloride is
$\Delta H^{\ominus} = 2326 - 1920 - 728 = -322\,kJ\,mol^{-1}$

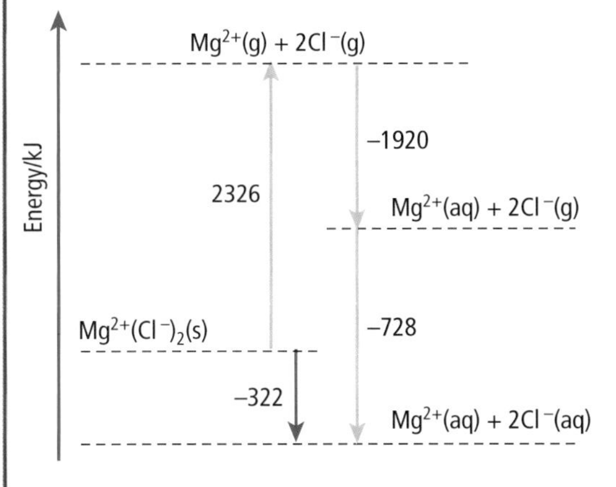

The diagram shows the thermochemical cycle used to calculate the enthalpy of solution of magnesium chloride.

DON'T FORGET

Values of enthalpy changes when attractions are broken are positive, since energy is required to break bonds or attractions. Values of enthalpy changes when attractions formed are negative, since energy is given out when attractions or bonds are formed.

LET'S THINK ABOUT THIS

1. The electron affinities for the elements forming ions with a –1 charge, given in the SQA Data Booklet, all have negative enthalpy values. Suggest why.
2. Suggest why the electron affinity for $O^-(g) + e^- \rightarrow O^{2-}(g)$ has a large positive value (+844 kJ).
3. Calculate the enthalpy of formation for $Na^+Cl^-(s)$ using enthalpy values from the SQA Data Booklet for the different steps. It is your choice whether to do this as a Born–Haber cycle.
4. Draw out a thermochemical cycle with enthalpy values to calculate the enthalpy of solution of calcium chloride.

REACTION FEASIBILITY 1

ENTROPY

The **entropy** of a system is a measure of the disorder within that system – the larger the entropy, the greater the disorder. Entropy is given the symbol **S** and the standard entropy of a substance, **$S^{\ominus}$**, is the entropy of one mole of the substance at a pressure of one atmosphere and normally at a temperature of 298 K. $S^{\ominus}$ values for a number of selected substances are given on page 16 of the SQA Data Booklet. Notice that the units of entropy are **$J\,K^{-1}\,mol^{-1}$**.

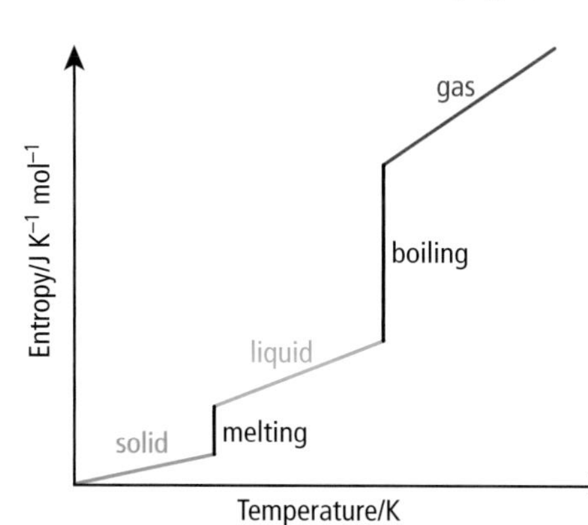

The graph shows how the entropy of a substance varies with temperature.

Substances in the solid state tend to have low entropy values. This is not surprising since the particles in a solid occupy approximately fixed positions. They can vibrate but can't move from one place to another. Solids, therefore, are highly ordered. Gases, on the other hand, have very high entropy values. They contain particles which have complete freedom of movement and, as a result, they are highly disordered. The entropies of liquids lie between these two extremes.

At 0 K, the particles in a solid are no longer vibrating and are perfectly ordered. So the entropy of a substance at 0 K is zero. This is known as the **Third Law of Thermodynamics**. As temperature increases, the entropy of the solid increases until the melting point, where there is a rapid increase in entropy as the solid changes into a liquid. You can see that there is an even larger increase in entropy at the boiling point as the liquid changes into a gas.

The standard entropy change ($\Delta S^{\ominus}$) in a chemical reaction can be calculated using the expression:

$\Delta S^{\ominus} = \sum S^{\ominus}$ (products) – $\sum S^{\ominus}$ (reactants)

When calculating $\Delta S^{\ominus}$ for a reaction, make sure you multiply the standard entropy of each substance by its corresponding stoichiometric coefficient.

Worked example

Using information from the SQA Data Booklet and the fact that silver(I) nitrate and nitrogen dioxide have $S^{\ominus}$ values of 142 and 241 $J\,K^{-1}\,mol^{-1}$ respectively, calculate the standard entropy change for the decomposition of silver(I) nitrate:

$2AgNO_3(s) \rightarrow 2Ag(s) + 2NO_2(g) + O_2(g)$

Using the above expression, $\Delta S^{\ominus} = [2(43) + 2(241) + (205)] - [2(142)] = +489\,J\,K^{-1}\,mol^{-1}$

FREE ENERGY

Feasible reactions

A **feasible** reaction is one which takes place of its own accord. It is important to note that a reaction which is feasible does not necessarily imply that it is fast. The conversion of diamond into graphite, for example, is feasible but takes millions of years.

It is the **Second Law of Thermodynamics** that defines the conditions of a feasible reaction. It states that **for a reaction to be feasible, the total entropy change for a reaction system and its surroundings must be positive**, that is the total entropy must increase:

$\Delta S^{\ominus}_{(total)} = \Delta S^{\ominus}_{(surroundings)} + \Delta S^{\ominus}_{(system)} = +ve$

Let us consider the following feasible reaction and calculate $\Delta S^{\ominus}_{(total)}$:
$NH_3(g) + HCl(g) \rightarrow NH_4Cl(s)$

$\Delta S^{\ominus}_{(system)}$ for this reaction is $-284\,J\,K^{-1}\,mol^{-1}$; the reaction is exothermic ($\Delta H^{\ominus} = -176\,kJ\,mol^{-1}$). The heat energy leaving the system produces an increase in the entropy of the surroundings since hot surroundings have a higher entropy than cold surroundings.

$\Delta S^{\ominus}_{(surroundings)}$ can be calculated using the relationship, $\Delta S^{\ominus}_{(surroundings)} = \frac{-\Delta H^{\ominus}}{T}$ where T is the temperature, which we will take as standard temperature (298 K).

contd

FREE ENERGY contd

So, $\Delta S^{\ominus}_{(surroundings)} = \frac{-\Delta H^{\ominus}}{T} = \frac{-(-176)}{298} = 0{\cdot}591\ \text{kJ K}^{-1}\,\text{mol}^{-1} = +591\ \text{J K}^{-1}\,\text{mol}^{-1}$

Hence, $\Delta S^{\ominus}_{(total)} = \Delta S^{\ominus}_{(surroundings)} + \Delta S^{\ominus}_{(system)} = (+591) + (-284) = +307\ \text{J K}^{-1}\,\text{mol}^{-1}$

The overall entropy change is positive which confirms the reaction is feasible.

Standard free energy change

We know that: $\Delta S^{\ominus}_{(total)} = \Delta S^{\ominus}_{(surroundings)} + \Delta S^{\ominus}_{(system)}$ and $\Delta S^{\ominus}_{(surroundings)} = \frac{-\Delta H^{\ominus}}{T}$

Hence, $\Delta S^{\ominus}_{(total)} = \frac{-\Delta H^{\ominus}}{T} + \Delta S^{\ominus}_{(system)}$

Multiplying this expression throughout by –T, we obtain: $-T\Delta S^{\ominus}_{(total)} = \Delta H^{\ominus} - T\Delta S^{\ominus}_{(system)}$

The quantity, $-T\Delta S^{\ominus}_{(total)}$, has units of energy and we call this energy change, the **standard free energy change**. It is given the symbol $\mathbf{\Delta G^{\ominus}}$.

So, $\mathbf{\Delta G^{\ominus} = \Delta H^{\ominus} - T\Delta S^{\ominus}}$

Since $\Delta S^{\ominus}_{(total)}$ must be positive for a reaction to be feasible, it follows that $\Delta G^{\ominus}$ must be negative, $\Delta G^{\ominus} < 0$.

The above expression can be used to predict whether a reaction will be feasible or not. Consider the decomposition of sodium hydrogencarbonate:

$2NaHCO_3(s) \rightarrow Na_2CO_3(s) + CO_2(g) + H_2O(g)$

For this reaction, $\Delta H^{\ominus} = +129\ \text{kJ mol}^{-1}$ and $\Delta S^{\ominus} = +335\ \text{J K}^{-1}\,\text{mol}^{-1}$ or $+0{\cdot}335\ \text{kJ K}^{-1}\,\text{mol}^{-1}$.
Hence, at standard temperature (298 K), $\Delta G^{\ominus} = +129 - 298(0{\cdot}335) = +29\ \text{kJ mol}^{-1}$

Since $\Delta G^{\ominus}$ is positive, it can be concluded that the reaction is not feasible at 298 K.

A reaction is feasible when $\Delta G^{\ominus}$ is negative and so it just becomes feasible when $\mathbf{\Delta G^{\ominus} = 0}$. Substituting $\Delta G^{\ominus} = 0$ in the above expression gives:

$0 = \Delta H^{\ominus} - T\Delta S^{\ominus}$ which can be arranged to $T = \frac{\Delta H^{\ominus}}{\Delta S^{\ominus}}$

So, the decomposition of sodium hydrogencarbonate just becomes feasible at

$T = \frac{\Delta H^{\ominus}}{\Delta S^{\ominus}} = \frac{129}{0{\cdot}335} = 385\ \text{K}$ (or 112 °C)

In these calculations involving temperature (T), units of Kelvin (K) are used.

Just as the standard enthalpy change ($\Delta H^{\ominus}$) for a reaction can be calculated from the standard enthalpies of formation ($\Delta H^{\ominus}_f$) of the reactants and products (see page 42), the standard free energy change ($\Delta G^{\ominus}$) for a reaction can be calculated from the standard free energies of formation ($\Delta G^{\ominus}_f$) of the reactants and products using the following expression:

$\Delta G^{\ominus} = \sum G^{\ominus}_f(\text{products}) - \sum G^{\ominus}_f(\text{reactants})$

LET'S THINK ABOUT THIS

If you hold a wide rubber band against your upper lip and quickly stretch it to its limit, you should detect a slight rise in temperature. This means that stretching rubber is an exothermic process, but could this have been predicted? Unstretched rubber consists of a tangled mass of polymer molecules and, since it is highly disordered, it has a high entropy. As the rubber is stretched, the polymer molecules straighten out and become more ordered. The entropy decreases as a result. So, $\Delta S^{\ominus}$ must be negative. Stretching rubber does not take place of its own accord and so, $\Delta G^{\ominus}$ must be positive. If we rearrange the expression, $\Delta G^{\ominus} = \Delta H^{\ominus} - T\Delta S^{\ominus}$, to make $\Delta H^{\ominus}$ the subject, we obtain $\Delta H^{\ominus} = \Delta G^{\ominus} + T\Delta S^{\ominus}$. Since $\Delta G^{\ominus}$ is positive, T is positive and $\Delta S^{\ominus}$ is negative, it follows that $\Delta H^{\ominus}$ could be positive (endothermic) **or** negative (exothermic), depending on the magnitudes of $\Delta G^{\ominus}$ and $\Delta S^{\ominus}$.

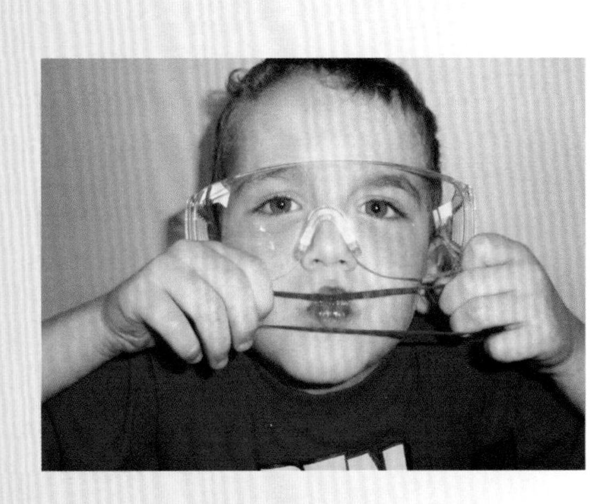

REACTION FEASIBILITY 2

FREE ENERGY contd

Free energy change and equilibrium

The standard free energy change ($\Delta G^\ominus$) for a reaction can give information about the equilibrium position in a reversible reaction and the value of K, the equilibrium constant.

If $\Delta G^\ominus < 0$ then the forward reaction will be feasible, so the products will predominate over the reactants; the equilibrium position will lie to the side of the products and K will be greater than 1 ($K > 1$).

If $\Delta G^\ominus > 0$ the reverse reaction will be feasible, so the reactants will predominate over the products; the equilibrium position will lie to the side of the reactants and K will be less than 1 ($K < 1$).

Consider the reversible reaction: $\mathbf{R} \rightleftharpoons \mathbf{P}$

Suppose $\Delta G^\ominus$ is negative, $\Delta G^\ominus = G^\ominus_p - G^\ominus_R = -ve$ where $G^\ominus_p$ and $G^\ominus_R$ are the standard free energies of the products and reactants respectively.

If we start with 1 mole of pure **R** at 1 atmosphere of pressure then standard state conditions apply and so, just before the reaction starts, we can talk about the standard free energy of **R** ($G^\ominus_R$) as opposed to the free energy of **R** (G_R).

However, as soon as the reaction starts and some **R** is converted into **P**, standard state conditions no longer apply. So, during a chemical reaction, we must talk about free energy (G) rather than standard free energy ($G^\ominus$).

As this reaction approaches equilibrium, the free energy of the system proceeds to a minimum as illustrated in the diagram. Since $\Delta G^\ominus$ is negative, the products predominate over the reactants, so the equilibrium position lies on the product side.

When equilibrium is established, the free energy of R (G_R) will equal the free energy of P (G_P).

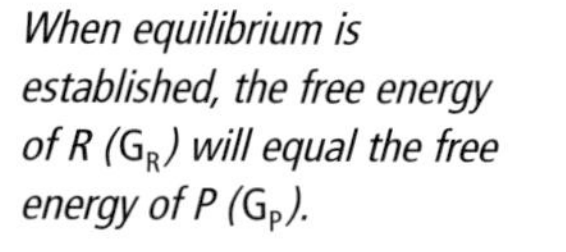

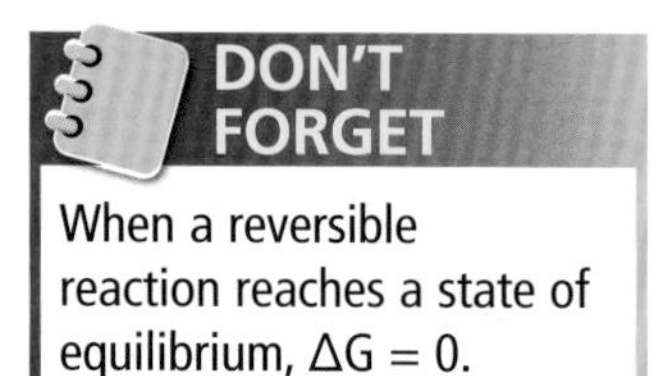
When a reversible reaction reaches a state of equilibrium, $\Delta G = 0$.

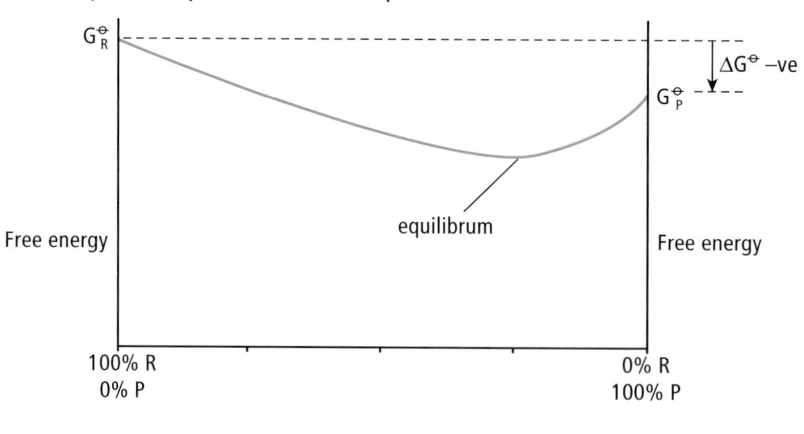

At equilibrium $\Delta G = G_P - G_R = 0$

It is important to note that at equilibrium, $\Delta G = 0$ and **not** $\Delta G^\ominus = 0$.

Ellingham diagrams

Consider the relationship: $\Delta G^\ominus = \Delta H^\ominus - T\Delta S^\ominus$

which can be rearranged to: $\Delta G^\ominus = -\Delta S^\ominus T + \Delta H^\ominus$

Comparing this with the equation for a straight line, $y = mx + c$, we see that a plot of standard free energy change ($\Delta G^\ominus$) against temperature (T) will have a gradient of $-\Delta S^\ominus$ and an intercept on the $\Delta G^\ominus$ axis of $\Delta H^\ominus$.

contd

FREE ENERGY contd

A plot of $\Delta G^\ominus$ against T for a reaction is known as an Ellingham diagram and the graph shows Ellingham diagrams for a few oxidation reactions.

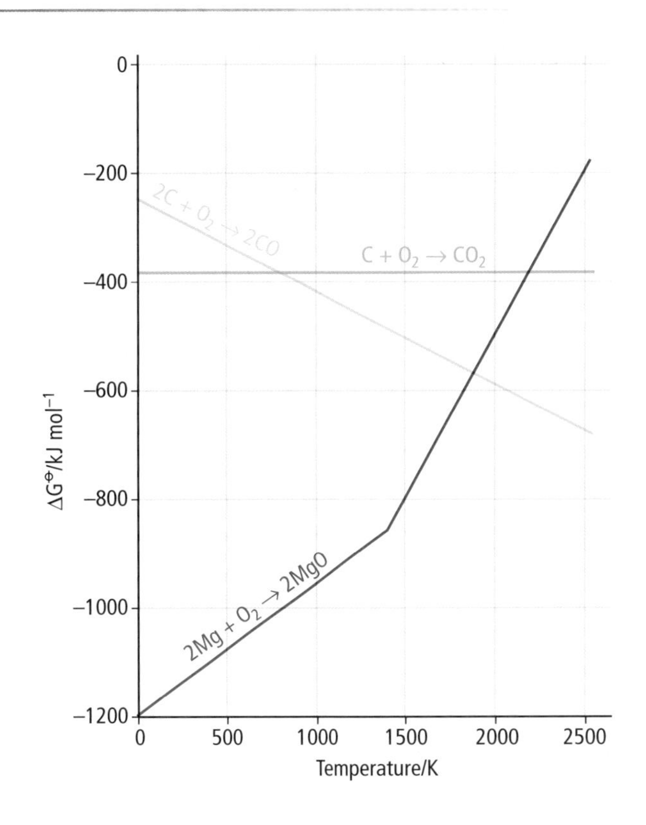

Consider the blue line which corresponds to the reaction, $2C(s) + O_2(g) \rightarrow 2CO(g)$

The entropy will increase in this reaction since **two** moles of CO(g) have a higher entropy than **one** mole of $O_2(g)$.

$\Delta S^\ominus$ is therefore positive and so the gradient of the line ($-\Delta S^\ominus$) will be negative. This is why the blue line slopes downwards.

Consider now the green line which corresponds to the reaction, $C(s) + O_2(g) \rightarrow CO_2(g)$

The entropy change here will be very small as $\Delta S^\ominus$ is approximately zero, and this is why the gradient of the green line is virtually zero.

You will have noticed that there is a break in the red line at approximately 1400 K: $2Mg + O_2 \rightarrow 2MgO$

This temperature corresponds to the boiling point of magnesium. As a result of the change in state of the magnesium, $\Delta S^\ominus$ for the reaction will change and, hence, the gradient of the red line will also change.

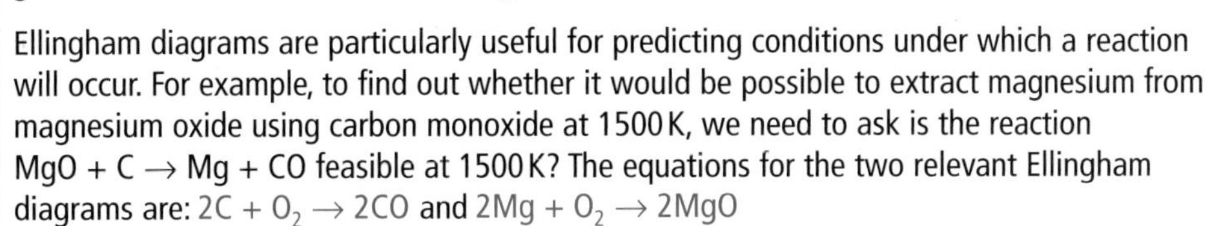

Ellingham diagrams are particularly useful for predicting conditions under which a reaction will occur. For example, to find out whether it would be possible to extract magnesium from magnesium oxide using carbon monoxide at 1500 K, we need to ask is the reaction $MgO + C \rightarrow Mg + CO$ feasible at 1500 K? The equations for the two relevant Ellingham diagrams are: $2C + O_2 \rightarrow 2CO$ and $2Mg + O_2 \rightarrow 2MgO$

However, we need the second reaction to go in reverse in order to achieve the desired overall reaction. From the graph it can be deduced that at 1500 K,
$2C + O_2 \rightarrow 2CO \quad \Delta G^\ominus = -500\,kJ\,mol^{-1} \quad 2MgO \rightarrow 2Mg + O_2 \quad \Delta G^\ominus = +800\,kJ\,mol^{-1}$

Notice that the sign of $\Delta G^\ominus$ for the second reaction is positive. This is because we need the reverse of the reaction shown in the Ellingham diagram. On adding these latter two equations we obtain: $2MgO + 2C \rightarrow 2Mg + 2CO \quad \Delta G^\ominus = +800 - 500 = +300\,kJ\,mol^{-1}$

Dividing throughout by 2, we obtain the desired equation and its $\Delta G^\ominus$ value:
$MgO + C \rightarrow Mg + CO \quad \Delta G^\ominus = +150\,kJ\,mol^{-1}$

Since $\Delta G^\ominus$ is positive, this reaction is not feasible at 1500 K. But at what temperature will it become feasible? You will have noticed that the two lines intersect at approximately 1900 K; it can be shown by calculation that $\Delta G^\ominus$ for the desired reaction at this temperature is zero, so the reaction just becomes feasible at 1900 K.

You can short circuit these calculations and come to a conclusion about the feasibility of a reaction purely by inspection of the relevant Ellingham diagrams. In general, for a pair of Ellingham diagrams, the reaction represented by the lower line will go as written, while the reaction for the upper line will go in reverse. This is because the reaction for the lower line has a more negative $\Delta G^\ominus$ value. Hence, carbon will reduce MgO to Mg at any temperature above approximately 1900 K.

DON'T FORGET

An Ellingham diagram shows how the standard free energy change ($\Delta G^\ominus$) for a reaction varies with temperature (T).

LET'S THINK ABOUT THIS

The graph on the opposite page shows the variation in free energy for the reversible reaction **R ⇌ P**. Use this graph to calculate an approximate value for K, the equilibrium constant for the reaction.

ELECTROCHEMISTRY

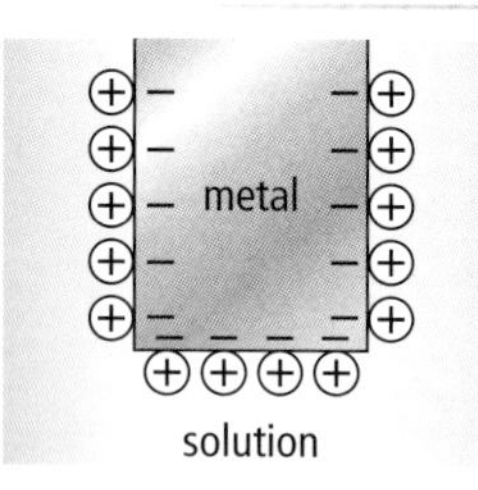

When a piece of metal is dipped into an ionic solution, the metal takes on a negative charge while the solution immediately surrounding it is positively charged.

ELECTRODE POTENTIALS, ELECTROCHEMICAL CELLS AND THE HYDROGEN HALF CELL

When a piece of metal is dipped into an ionic solution, there is a tendency for the metal to ionise. The metal ions that are formed move into the solution leaving electrons on the surface. The metal takes on a negative charge, while the solution immediately surrounding it is positively charged. A charge difference is therefore set up between the metal and the solution, and this is known as the **electrode potential**. It varies from one metal to another depending on the ease with which the metal ionises.

Consider this **electrochemical cell** in which zinc and silver half cells are linked. In this cell electrons will flow from zinc to silver through the external circuit. A **salt bridge** (or ion bridge) links the half cells internally and its purpose is to allow ions to flow from one half cell to the other. The bridge can be a piece of filter paper soaked in an aqueous solution of an ionic compound such as ammonium nitrate.

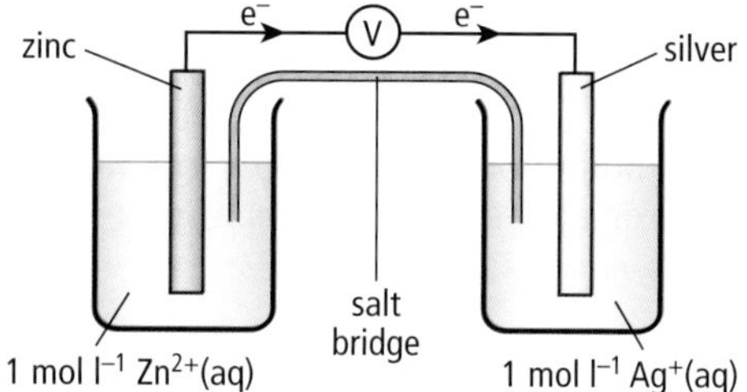

Zinc and silver half cells linked by a salt bridge to make an electrochemical cell.

This cell can be represented in abbreviated form as: $Zn(s)|Zn^{2+}(aq)||Ag^{+}(aq)|Ag(s)$ where $||$ represents the salt bridge.

When a voltmeter is placed in the external circuit, the voltage that it registers corresponds to the **difference** in the electrode potentials of the two half cells involved. Values of electrode potentials for a number of half cells are listed on page 11 of the SQA Data Booklet. These are **standard electrode potentials ($E^{\ominus}$)**, that is electrode potentials measured under standard conditions.

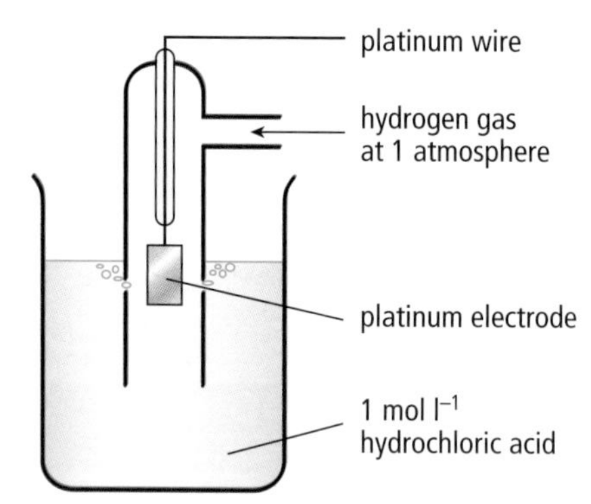

The standard hydrogen half cell is used as the reference half cell. It can be represented as $Pt(s)|H_2(g)|H^+(aq)$.

The **standard conditions** are:

- The concentration of the ions involved in the half reaction must be 1 mol l^{-1}.
- If a gas is involved, it must be at a pressure of 1 atmosphere.
- If the temperature is not specified, it is assumed to be 298 K.

Standard electrode potentials are often referred to as **standard reduction potentials** simply because the ion–electron equations listed on page 11 are conventionally written as reductions.

While the difference in the electrode potentials of two half cells can be measured, it is not possible to measure the electrode potential of a half cell on its own. To overcome this problem, a **reference half cell** (the standard hydrogen half cell) has been chosen and arbitrarily assigned an $E^{\ominus}$ value of 0·00 V.

The standard electrode potential of a half cell can be determined by linking it with a standard hydrogen half cell and measuring the voltage of the cell that is formed.

In order to determine the standard electrode potential of a half cell, an electrochemical cell is set up linking this half cell to a standard hydrogen half cell. The emf or cell voltage is then measured. This will equal the standard electrode potential of the half cell under test, since the $E^{\ominus}$ value for the standard hydrogen half cell is 0·00 V.

When measuring cell voltages, it is important to use a high resistance voltmeter since the voltage of a cell is defined as the potential difference between the electrodes when no current is drawn.

CALCULATING CELL VOLTAGES

Consider the electrochemical cell, $Ag(s) | Ag^{+}(aq) || Cl^{-}(aq) | Cl_2(g) | Pt(s)$ operating under standard conditions. Electrons will flow in the external circuit from silver to chlorine through the platinum and the following half reactions will take place at the electrodes:

$Ag(s) \rightarrow Ag^{+}(aq) + e^{-}$ $E^{\ominus} = -0{\cdot}80\,V$ $Cl_2(g) + 2e^{-} \rightarrow 2Cl^{-}(aq)$ $E^{\ominus} = +1{\cdot}36\,V$

contd

CALCULATING CELL VOLTAGES contd

The sign of silver's $E^{\ominus}$ value has been changed from that given in the SQA Data Booklet since we are dealing here with an oxidation potential rather than a reduction potential. To calculate the cell voltage, you simply add the $E^{\ominus}$ values together, cell voltage = $(-0{\cdot}80) + (+1{\cdot}36) = 0{\cdot}56\,V$

PREDICTING THE RELATIVE STRENGTHS OF OXIDISING AND REDUCING AGENTS

An oxidising agent is an electron acceptor and a reducing agent is an electron donor; the relative strengths of oxidising and reducing agents can be predicted from their corresponding $E^{\ominus}$ values. The $E^{\ominus}$ value is a measure of how readily a species is reduced. The more positive the $E^{\ominus}$ value, the more readily it will be reduced and the better it will be as an oxidising agent. For example, tin(IV) ions ($E^{\ominus} = +0{\cdot}15\,V$) will be stronger oxidising agents than tin(II) ions ($E^{\ominus} = -0{\cdot}14\,V$).

So, the best oxidising agents are to be found at the bottom of the table of reduction potentials (see page 11 of the SQA Data Booklet) and on the left-hand side. This implies that the best reducing agents are to be found at the top of the table on the right-hand side. As you descend the table on the right-hand side, the strength of the reducing agent decreases. So, for example, silver atoms will be weaker oxidising agents than zinc atoms.

STANDARD FREE ENERGY CHANGE AND CELL VOLTAGE

For a cell operating under standard conditions, the standard free energy change ($\Delta G^{\ominus}$) for the cell reaction is related to the cell voltage or emf ($E^{\ominus}$) by the expression: $\Delta G^{\ominus} = -nFE^{\ominus}$ where **n** is the number of moles of electrons transferred in the cell reaction and **F** is the Faraday constant ($9{\cdot}65 \times 10^4\,C\,mol^{-1}$).

Consider again the Ag/Cl_2 cell. Under standard conditions, we calculated its cell voltage to be $+0{\cdot}56\,V$. The ion-electron equations for the half reactions taking place are:
$2Ag(s) \rightarrow 2Ag^+(aq) + 2e^-$ and $Cl_2(g) + 2e^- \rightarrow 2Cl^-(aq)$
and the overall cell reaction is: $2Ag(s) + Cl_2(g) \rightarrow 2Ag^+(aq) + 2Cl^-(aq)$

Notice that the oxidation ion-electron equation has been doubled in order to obtain the balanced redox equation. For 1 mole of this redox reaction, 2 moles of electrons have been transferred from the silver atoms to the chlorine molecules and so n takes the value 2.

Hence, $\Delta G^{\ominus} = -nFE^{\ominus} = -2 \times 9{\cdot}65 \times 10^4 \times 0{\cdot}56 = -1{\cdot}08 \times 10^5\,J\,mol^{-1} = -108\,kJ\,mol^{-1}$

For a reaction to be feasible, we know that $\Delta G^{\ominus}$ must be negative. It therefore follows from the expression, $\Delta G^{\ominus} = -nFE^{\ominus}$, that a cell reaction can only be feasible if the cell voltage ($E^{\ominus}$) is positive.

DON'T FORGET

It is vital that you remember the negative sign when using the expression $\Delta G^{\ominus} = -nFE^{\ominus}$.

FUEL CELLS

A **fuel cell** is like any other electrochemical cell in that it generates electricity directly from a redox reaction. However, the reactants are continually being added so a fuel cell should continue to work indefinitely. The simplest fuel cell is the **hydrogen fuel cell** shown in the diagram.

At the Ni electrode on the left, the hydrogen reacts with hydroxide ions:
$H_2(g) + 2OH^-(aq) \rightarrow 2H_2O(l) + 2e^-$ $E^{\ominus} = +0{\cdot}83\,V$

At the Ni electrode on the right, the oxygen reacts with water:
$O_2(g) + 2H_2O(l) + 4e^- \rightarrow 4OH^-(aq)$ $E^{\ominus} = +0{\cdot}40\,V$

The overall cell reaction is: $2H_2(g) + O_2(g) \rightarrow 2H_2O(l)$
and the cell voltage under standard conditions takes the value $+1{\cdot}23\,V$.

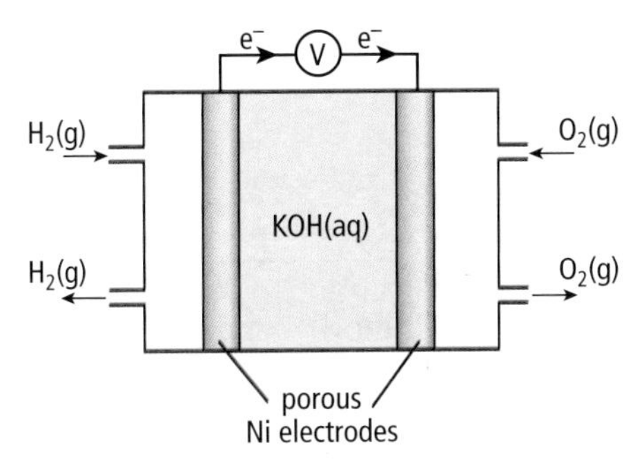

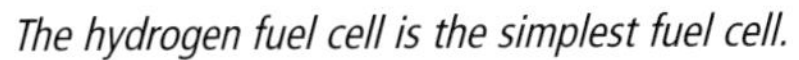

The hydrogen fuel cell is the simplest fuel cell.

LET'S THINK ABOUT THIS

Calculate the standard free energy change for the cell reaction that takes place in the hydrogen fuel cell described above.

KINETICS

RATE EQUATION, ORDER OF REACTION AND RATE CONSTANT

The rate of a chemical reaction normally depends on the concentrations of the reactants. Consider the reaction: **A** + **B** $\rightarrow$ products

Suppose, when the initial concentration of **A** is doubled and the initial concentration of **B** is kept constant, the rate of the reaction doubles. This implies that the rate of reaction is directly proportional to the concentration of **A**, so rate $\alpha\,[\mathbf{A}]^1$.

Suppose too, that the rate increases by a factor of four when the initial concentration of **B** is doubled and that of **A** is kept constant. This implies that the rate is directly proportional to the square of the concentration of **B**, so rate $\alpha\,[\mathbf{B}]^2$.

Combining these results gives: rate $\alpha\,[\mathbf{A}]^1[\mathbf{B}]^2 = k[\mathbf{A}]^1[\mathbf{B}]^2$............(1)
where k is a constant. We say that the reaction is **first order** with respect to **A** and **second order** with respect to **B**.

In general, for a reaction of the type a**A** + b**B** $\rightarrow$ products, the rate's dependence on the concentrations of A and B may be expressed in the following way: rate = $k[\mathbf{A}]^m[\mathbf{B}]^n$

DON'T FORGET

There is no relationship between orders and stoichiometric coefficients.

An expression of this kind is known as the **rate equation** for the reaction. The indices, m and n, are the **orders of the reaction** with respect to **A** and **B** respectively and they bear no relationship to the stoichiometric coefficients, a and b. The reaction is said to be m order with respect to **A** and n order with respect to **B**. The orders m and n are usually small whole numbers and rarely greater than 2.

The **overall order of the reaction** is given by (m + n). For example, in the reaction described by rate equation (1) above, the overall order would be 3 (1 + 2).

The constant **k** in the rate equation is known as the **rate constant** and the **units of rate constant** vary depending on the overall order of the reaction. Suppose a reaction is third order overall. Its rate equation could take the form: rate = $k[A]^2[B]^1$

Hence, $k = \dfrac{\text{rate}}{[A]^2[B]^1}$

Rate can be measured in $\text{mol l}^{-1}\text{s}^{-1}$ and the units of concentration are mol l^{-1}, so the units for k for this third order reaction are $\dfrac{\text{mol l}^{-1}\text{s}^{-1}}{(\text{mol}^2\text{l}^{-2})(\text{mol l}^{-1})} = \text{mol}^{-2}\text{l}^2\text{s}^{-1}$

Rate constant units for zero to third order reactions are summarised in the table.

Overall order	Units of k
0	$\text{mol l}^{-1}\text{s}^{-1}$
1	s^{-1}
2	$\text{mol}^{-1}\text{l s}^{-1}$
3	$\text{mol}^{-2}\text{l}^2\text{s}^{-1}$

DETERMINING RATE EQUATIONS AND RATE CONSTANTS

The rate equation for a chemical reaction can only be derived experimentally. A series of experiments in which the initial concentrations of the reactants are varied has to be carried out. The initial rate for each experiment is determined.

Consider the following experimental data for the reaction: A + B + C $\rightarrow$ products

Experiment	[A]/ mol l^{-1}	[B]/ mol l^{-1}	[C]/ mol l^{-1}	Initial rate/ $\text{mol l}^{-1}\text{s}^{-1}$
1	1·0	1·0	1·0	20
2	2·0	1·0	1·0	40
3	1·0	2·0	1·0	20
4	1·0	1·0	2·0	80

By comparing experiments 1 and 2, you can see that doubling the concentration of A increases the reaction rate by a factor of two, which implies the reaction is first order with respect to A.

Comparing experiments 1 and 3 shows that doubling the concentration of B has no effect on the rate. This means the reaction is zero order with respect to B.

Comparing experiments 1 and 4 indicates that doubling the concentration of C increases the reaction rate by a factor of four. This implies the reaction is second order with respect to C.

contd

DETERMINING RATE EQUATIONS AND RATE CONSTANTS contd

Hence the rate equation is: rate = $k[A]^1[B]^0[C]^2$ or more simply, rate = $k[A]^1[C]^2$

The reaction is third order overall.

The rate constant can be determined by substituting experimental data from any one of the four experiments into the rearranged rate equation. Let's take data from experiment 2.

$$k = \frac{rate}{[A]^1[C]^2} = \frac{40}{(2.0) \times (1.0)^2} = 20\,mol^{-2}\,l^2\,s^{-1}$$

REACTION MECHANISM

Most reactions are believed to proceed by a series of steps rather than by one single step. This series of steps is known as the **reaction mechanism**. The overall rate of a reaction is dependent on the slowest step which is called the **rate-determining step**.

Once the kinetics of a reaction have been worked out, it is possible to propose a mechanism for the reaction. Consider, for example, the reaction between nitrogen dioxide and fluorine. The stoichiometric equation for the reaction is: $2NO_2 + F_2 \rightarrow 2NO_2F$ and the experimentally determined rate equation is: rate = $k[NO_2][F_2]$

The reaction is **first** order with respect to each of the reactants and this implies that **one** molecule of NO_2 and **one** molecule of F_2 must be involved in the slow, rate-determining step. The following reaction mechanism can be proposed:

step 1 $NO_2 + F_2 \rightarrow X$ *slow*
step 2 $NO_2 + X \rightarrow 2NO_2F$ *fast*

From additional experimental evidence, it has been deduced that intermediate X is a mixture of NO_2F and F. The mechanism for the reaction is therefore:

step 1 $NO_2 + F_2 \rightarrow NO_2F + F$ *slow*
step 2 $NO_2 + NO_2F + F \rightarrow 2NO_2F$ *fast*

Adding the two steps together gives: $2NO_2 + F_2 \rightarrow 2NO_2F$ which is and must be identical to the stoichiometric equation for the mechanism to be valid.

It is important to note that an experimentally determined rate equation can provide evidence for a proposed reaction mechanism, but it cannot provide proof since other possible reaction mechanisms may give the same rate equation.

DON'T FORGET

A rate equation can **only** tell us which species react together in the rate-determining step and how many particles of each are involved in that step.

LET'S THINK ABOUT THIS

Bromomethane undergoes a substitution reaction with sodium hydroxide solution.

$CH_3Br + OH^- \rightarrow CH_3OH + Br^-$

The experimentally determined rate equation is rate = $k[CH_3Br][OH^-]$

The table shows some kinetic data for this reaction.

Experiment	$[CH_3Br]/mol\,l^{-1}$	$[OH^-]/mol\,l^{-1}$	Relative rate
1	0·05	0·10	1
2	0·05	0·20	*x*
3	0·10	*y*	10

Determine values for ***x*** and ***y***.

UNIT 2 PPAs 1–5

UNIT 2 PPA 1 – COMPLEXOMETRIC DETERMINATION OF NICKEL USING EDTA

The EDTA–nickel complex.

Aim

To determine the percentage of nickel in a nickel salt using a complexometric technique.

Introduction

This PPA is a special case of volumetric analysis known as complexometric analysis. EDTA is a hexadentate ligand which binds in a 1 : 1 ratio with most metal ions, for example Ni^{2+}, forming a stable octahedral **complex**. The coordination number of the central nickel ion in the complex is 6.

An ordinary indicator cannot be used since the reaction does not involve a simple acid–alkali neutralisation. The indicator used is murexide which is a different colour when free, compared to its colour when it is attached to Ni^{2+} ions. Murexide is a suitable indicator since it binds less strongly to the Ni^{2+} ions than does EDTA.

Procedure

DON'T FORGET

The indicator used in this titration is murexide and you must know that it is one colour when attached to Ni^{2+} ions and a different colour when unattached. It binds less strongly to the Ni^{2+} ions than does EDTA, and at the end-point the murexide is no longer attached to the Ni^{2+} ions.

In the experiment approximately 2·6 g of hydrated nickel(II) sulphate was weighed out accurately and dissolved in about 25 cm^3 of deionised water. This was transferred with rinsings into a 100 cm^3 standard flask and made up to the mark with more deionised water. The flask was then shaken thoroughly to ensure that the contents were well mixed.

20 cm^3 of this solution was pipetted into a conical flask. A small quantity of murexide indicator was added along with approximately 10 cm^3 of ammonium chloride solution.

0·10 mol l^{-1} EDTA solution was dispensed from the burette and approximately 10 cm^3 of concentrated ammonia solution was added into the conical flask near the end-point. The colour changed to blue–violet at the end-point.

The titrations were repeated until two concordant results were achieved.

Results

Theoretical calculation

FM of hydrated nickel(II) sulphate, $NiSO_4.6H_2O$ = 262·8

$\% \text{ Ni} = \frac{58{\cdot}7}{262{\cdot}8} \times 100 = 22{\cdot}3\%$

Experimental results

Mass of hydrated nickel(II) sulphate used = 2·63 g

	Rough titre	First titre	Second titre
Initial burette reading (cm^3)	0·1	0·2	20·4
Final burette reading (cm^3)	20·8	20·4	40·5
Volume of EDTA added (cm^3)	20·7	20·2	20·1

DON'T FORGET

EDTA reacts with nickel ions in a 1 : 1 ratio.

Average of concordant results $= \frac{20{\cdot}2 + 20{\cdot}1}{2} = 20{\cdot}15\ cm^3$

Concentration of EDTA solution = 0·10 mol l^{-1}

The number of moles of EDTA used to react with 20 cm^3 of nickel(II) solution is
$V \times c = 0{\cdot}02015 \times 0{\cdot}10 = 2{\cdot}015 \times 10^{-3}$ mol

contd

UNIT 2 PPA 1 – COMPLEXOMETRIC DETERMINATION OF NICKEL USING EDTA contd

Therefore, number of moles of Ni^{2+} ions in $20\,cm^3 = 2{\cdot}015 \times 10^{-3}$ mol

Number of moles of Ni^{2+} ions in the total volume of $100\,cm^3$
$= 2{\cdot}015 \times 10^{-3} \times 5 = 1{\cdot}0075 \times 10^{-2}$ mol

Therefore, mass of nickel in salt = $n \times FM = 1{\cdot}0075 \times 10^{-2} \times 58{\cdot}7 = 0{\cdot}5914$ g

From these results the % mass of nickel in the salt = $\frac{0{\cdot}5914}{2{\cdot}63} \times 100 = 22{\cdot}5\%$

Conclusion

The percentage by mass of nickel in hydrated nickel(II) sulphate was shown by experiment to be 22·5%. This compares very well with the theoretical value of 22·3%.

Evaluation

The experimental result fits in very well with the theoretical result. This suggests that the hydrated nickel(II) sulphate used was very pure. Impurities such as other metal ions would have reacted with the EDTA to give higher titre values and, therefore, a higher experimental value for the percentage of nickel present.

At first, it was quite difficult to decide when the end-point had been reached, but the appearance of the blue–violet colour was sharp. Note that in this reaction the ammonia–ammonium chloride solutions are used as a buffer to keep the pH constant. Many complexometric indicators and titrants are pH dependent.

Possible uncertainty values in the measurements include:

- burette readings $\pm$ 0·05 cm^3
- pipette volumes $\pm$ 0·06 cm^3
- balance readings $\pm$ 0·01 g
  (This means that a tared mass will have an uncertainty value of $\pm$ 0·02 g.)
- There is also an uncertainty in the concentration of the EDTA solution.

Uncertainty calculation

Uncertainty in mass of $NiSO_4.6H_2O = 0{\cdot}02$ g

% uncertainty in mass of $NiSO_4.6H_2O = \frac{0{\cdot}02}{2{\cdot}63} \times 100 = 0{\cdot}76\%$

Uncertainty in volume of $NiSO_4$(aq) ($100\,cm^3$ volumetric flask) = $0{\cdot}2\,cm^3$

% uncertainty in volume of $NiSO_4$(aq) = $\frac{0{\cdot}2}{100{\cdot}0} \times 100 = 0{\cdot}20\%$

Uncertainty in pipetted volume of $NiSO_4$(aq) ($20\,cm^3$ pipette) = $0{\cdot}06\,cm^3$

% uncertainty in pipetted volume of $NiSO_4$(aq) = $\frac{0{\cdot}06}{20{\cdot}0} \times 100 = 0{\cdot}30\%$

Uncertainty in concentration of EDTA = $0{\cdot}0002\,mol\,l^{-1}$
(this value would be provided by your teacher)

% uncertainty in concentration of EDTA = $\frac{0{\cdot}0002}{0{\cdot}100} \times 100 = 0{\cdot}20\%$

Uncertainty in titre volume of EDTA ($50\,cm^3$ burette) = $0.1\,cm^3$

% uncertainty in titre volume of EDTA = $\frac{0{\cdot}1}{20{\cdot}15} \times 100 = 0{\cdot}50\%$

% uncertainty in percentage of Ni = $0{\cdot}76 + 0{\cdot}20 + 0{\cdot}30 + 0{\cdot}20 + 0{\cdot}50 = 1{\cdot}96\,\%$

Absolute uncertainty in percentage of Ni = $\frac{1{\cdot}96}{100} \times 22{\cdot}5 = 0{\cdot}44\%$

Hence, percentage of Ni in $NiSO_4.6H_2O = 22{\cdot}5 \pm 0{\cdot}4\%$

For more information on uncertainty calculations see A Guide to Practical Work written by David Hawley and published by Learning and Teaching Scotland.

UNIT 2 PPA 2 – GRAVIMETRIC DETERMINATION OF WATER IN HYDRATED BARIUM CHLORIDE

Aim

To determine the value of n in the formula $BaCl_2.nH_2O$ using gravimetric analysis.

Gravimetric analysis involves accurately measuring the mass of a product; volumetric analysis involves accurately measuring volumes of reacting liquids.

Introduction

This is a simple example of gravimetric analysis. This technique involves weighing accurately rather than doing titrations. The hydrated barium chloride is heated until all the water has been driven off as in the equation: $BaCl_2.nH_2O \rightarrow BaCl_2 + nH_2O$

From the masses determined in the experiment, it is possible to calculate the relative number of moles of barium chloride and water, and thus calculate a value for n which must be a whole number.

Procedure

A crucible was heated to drive off any residual water and, after cooling in a desiccator, it was weighed accurately. Approximately 2·5 g of the hydrated barium chloride was added to the dry crucible and the crucible plus contents were reweighed, again accurately. From this, the accurate mass of the hydrated barium chloride can be calculated.

A desiccator provides a very dry atmosphere to allow the heated barium chloride to cool without reabsorbing any water.

The hydrated barium chloride was heated in the crucible using a blue Bunsen flame. The crucible plus contents were allowed to cool in a desiccator. The desiccator contains a drying agent which removes any moisture from the air. This prevents water being absorbed by the crucible and contents when cooling in the desiccator. The cooled crucible plus contents were then reweighed.

The heating, cooling in the desiccator, weighing cycle was repeated until a constant mass had been achieved. At this stage, it was presumed that all the water had been driven off.

Results

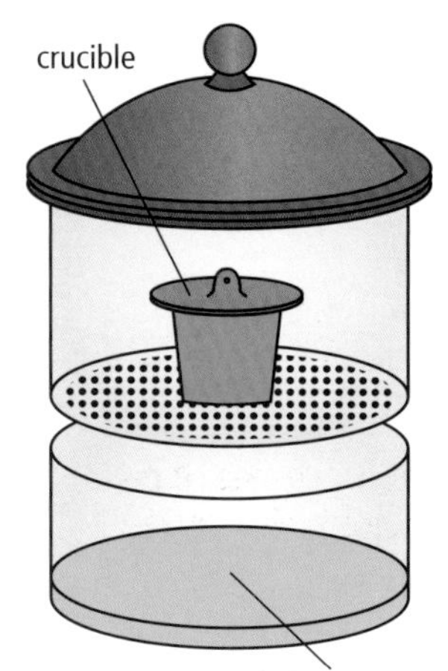

Mass of empty crucible = 32·67 g

Mass of crucible + hydrated barium chloride = 35·03 g

Therefore mass of hydrated barium chloride = 35·03 – 32·67 g = 2·36 g

Mass of crucible + anhydrous barium chloride = 34·69 g
(after constant mass had been reached)

Therefore, mass of anhydrous barium chloride = 34·69 – 32·67 = 2·02 g

Mass of water driven off = 2·36 – 2·02 = 0·34 g

Number of moles of $BaCl_2$, $n = \frac{\text{mass}}{\text{FM}} = \frac{2{\cdot}02}{208{\cdot}3} = 0{\cdot}00970\,\text{mol}$

Number of moles of water, $n \frac{\text{mass}}{\text{FM}} = \frac{0{\cdot}34}{18} = 0{\cdot}0189\,\text{mol}$

Ratio of moles of $BaCl_2 : H_2O$ = 0·00970 : 0·0189 = 1 : 1·95 which is approximately 1 : 2.

Conclusion

The formula of hydrated barium chloride is $BaCl_2.2H_2O$ (n = 2).

Evaluation

The fact that the experimental result fits the actual formula of hydrated barium chloride suggests that the technique is accurate and that the sample of hydrated barium chloride is pure.

Reasons for an inaccurate result include:

- errors in balance readings
- impurities present in the sample of hydrated barium chloride
- not all the water being driven off
- using a yellow flame instead of a blue flame when heating the crucible. (A yellow flame would result in deposits of soot being left on the base of the crucible, affecting the masses measured.)

UNIT 2 PPA 3 – DETERMINATION OF A PARTITION COEFFICIENT

Aim

To determine the partition coefficient when iodine distributes itself between aqueous potassium iodide and cyclohexane.

Introduction

When iodine is added to a pair of immiscible liquids, such as aqueous potassium iodide and cyclohexane, it distributes or partitions itself between the two liquids and the following equilibrium is established:

$I_2(aq) \rightleftharpoons I_2(C_6H_{12})$

The partition of the iodine between both liquids can be described quantitatively in terms of a partition coefficient. The partition coefficient is an example of an equilibrium constant, K, and the relevant expression is

$$K = \frac{[I_2(C_6H_{12})]}{[I_2(aq)]}$$

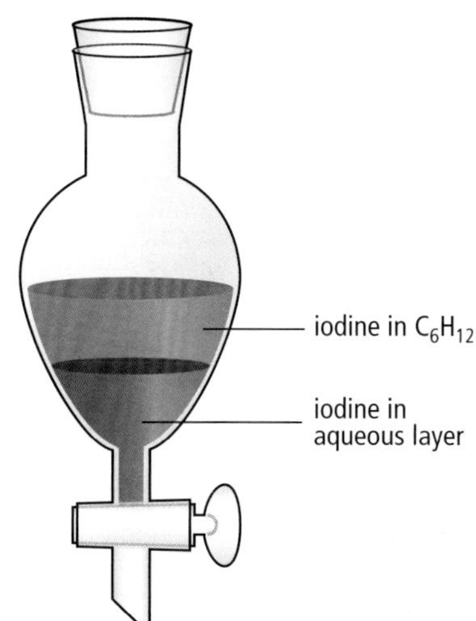

Procedure

50 cm^3 of 0·050 mol l^{-1} iodine solution (iodine dissolved in aqueous potassium iodide) was pipetted into a separating funnel. 50 cm^3 of cyclohexane was then pipetted into the same separating funnel. The separating funnel was stoppered and then shaken vigorously for about 2 minutes. The contents were allowed to settle and two layers separated out. The upper layer, which contained iodine in cyclohexane, was a purple colour and the lower layer which contained iodine in the aqueous potassium iodide was a reddish-brown colour.

To determine the concentration of iodine in the aqueous solution, the lower layer (the aqueous layer) was run off into a dry beaker and 10 cm^3 of this was pipetted into a conical flask. This was titrated against 0·050 mol l^{-1} sodium thiosulphate solution. During the titration, the brown colour of the iodine solution became lighter. When it became 'straw' coloured, a few drops of starch indicator solution were added. The colour of the solution became blue–black and the end-point of the titration was indicated when the solution in the conical flask became colourless. The titrations were repeated until two concordant results were obtained.

To determine the concentration of iodine in the cyclohexane, the upper cyclohexane layer was run off into a dry beaker. 10 cm^3 of this solution was pipetted into a conical flask and approximately 10 cm^3 of deionised water was added. This solution was then titrated against 0·025 mol l^{-1} sodium thiosulphate solution. A few drops of starch indicator solution were again added near the end-point. Since the starch solution and the cyclohexane layer did not mix well, it was necessary to stop the titration at intervals and shake the conical flask vigorously. The end-point of the titration was once again indicated when the solution in the conical flask became colourless. The titration was repeated until two concordant results were obtained.

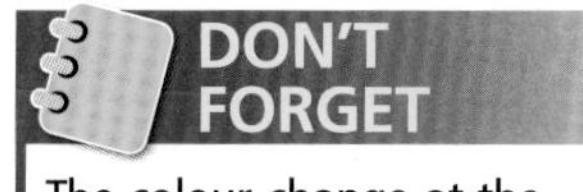

The colour change at the end-point is blue–black to colourless.

Results

Lower aqueous layer

Concentration of sodium thiosulphate solution = 0·050 mol l^{-1}

	Rough titration	First titration	Second titration
Initial burette reading (cm^3)	0·10	13·20	26·20
Final burette reading (cm^3)	13·20	26·00	39·05
Titre (cm^3)	13·10	12·80	12·85

Average of concordant results = 12·83 cm^3

contd

UNIT 2 PPA 3 – DETERMINATION OF A PARTITION COEFFICIENT contd

The number of moles of sodium thiosulphate used,
$n_{thiosulphate} = V \times c = 0{\cdot}01283 \times 0{\cdot}050 = 6{\cdot}415 \times 10^{-4}$ mol

The relevant ion-electron equations are:
$2S_2O_3^{2-} \rightarrow S_4O_6^{2-} + 2e^-$ and $I_2 + 2e^- \rightarrow 2I^-$

This tells us that 2 mol of $S_2O_3^{2-}$ reacts with 1 mol of I_2.

Therefore, the number of moles of iodine in the 10 cm³ sample pipetted from the aqueous layer is
$\frac{6{\cdot}415 \times 10^{-4}}{2} = 3{\cdot}21 \times 10^{-4}$ mol in 10 cm³ which is $3{\cdot}21 \times 10^{-2}$ mol l⁻¹

Upper cyclohexane layer

Concentration of sodium thiosulphate solution = 0·025 mol l^{-1}

	Rough titration	First titration	Second titration
Initial burette reading (cm³)	0·20	11·80	23·00
Final burette reading (cm³)	11·80	23·00	34·20
Titre (cm³)	11·60	11·20	11·20

Average of concordant results = 11·20 cm³

The number of moles of sodium thiosulphate used,
$n_{thiosulphate} = V \times c = 0{\cdot}01120 \times 0{\cdot}025 = 2{\cdot}80 \times 10^{-4}$ mol

Therefore, the number of moles of iodine in the 10 cm³ sample pipetted from the cyclohexane layer is
$\frac{2{\cdot}80 \times 10^{-4}}{2} = 1{\cdot}40 \times 10^{-4}$ mol in 10 cm³ which is $1{\cdot}40 \times 10^{-2}$ mol l^{-1}

$$K = \frac{[I_2(C_6H_{12})]}{[I_2(aq)]} = \frac{1{\cdot}40 \times 10^{-2}}{3{\cdot}21 \times 10^{-2}} = 0{\cdot}44$$

Conclusion

The partition coefficient when iodine distributes itself between aqueous potassium iodide and cyclohexane is 0·44.

The partition coefficient value has no units since it is an example of an equilibrium constant.

Evaluation

- A factor which may have affected the result was whether the distribution of iodine between the two solvents had reached equilibrium before the two layers were separated.
- Achieving the exact end-point point is always difficult in a titration, but the colour change was made easier to see using starch indicator. However, neither the starch solution nor the sodium thiosulphate solution mix well with cyclohexane. So, it was necessary to stop titrating and shake the contents of the conical flask regularly to ensure better mixing. This may have led to an error in getting the exact end-point.
- The calculation required the exact concentrations of the two sodium thiosulphate solutions used as titrants. If these were inaccurate, the partition coefficient would have been calculated incorrectly.
- Ideally, the whole experiment should have been repeated to make the result more reliable. The three titrations shown are not duplicates but repeated titrations to achieve concordant results.

UNIT 2 PPA 4 – VERIFICATION OF A THERMODYNAMIC PREDICTION

Aim

To calculate the theoretical temperature at which sodium hydrogencarbonate decomposes and to verify this experimentally.

Introduction

Sodium hydrogencarbonate, $NaHCO_3$, decomposes on heating to produce sodium carbonate, steam and carbon dioxide.

By carrying out calculations using the relevant thermodynamic data, it is possible to find the theoretical decomposition temperature and then carry out an experiment to verify this.

The relevant data is shown in the table.

Compound	Standard enthalpy of formation/kJ mol^{-1}	Standard entropy/J K^{-1} mol^{-1}
$NaHCO_3(s)$	–948	102
$Na_2CO_3(s)$	–1131	136
$H_2O(g)$	–242	189
$CO_2(g)$	–394	214

Using the following balanced equation and the above data, $\Delta H^{\ominus}$ and $\Delta S^{\ominus}$ for the reaction can be calculated as shown.

$$2NaHCO_3(s) \rightarrow Na_2CO_3(s) + H_2O(g) + CO_2(g)$$

$\Delta H_f^{\ominus}$ 2×-948 -1131 -242 -394

$S^{\ominus}$ 2×102 136 189 214

$\Delta H^{\ominus} = \Sigma \Delta H_f^{\ominus}(\text{products}) - \Sigma \Delta H_f^{\ominus}(\text{reactants})$

$= (-1131 + -242 + -394) - (2 \times -948) = 129\,\text{kJ mol}^{-1}$

$\Delta S^{\ominus} = \Sigma S^{\ominus}(\text{products}) - \Sigma S^{\ominus}(\text{reactants})$

$= (136 + 189 + 214) - (2 \times 102) = 335\,\text{J K}^{-1}\,\text{mol}^{-1}$

$\Delta G^{\ominus} = \Delta H^{\ominus} - T\Delta S^{\ominus} = 0$ at the temperature at which the reaction just becomes feasible.

Rearranging the equation gives, $T = \frac{\Delta H^{\ominus}}{\Delta S^{\ominus}} = \frac{129 \times 1000}{335} = 385\,\text{K} = 112°\text{C}$

Therefore, the theoretical decomposition temperature is 112°C.

You must be able to do calculations like this as similar ones are often in the SQA exam.

The experiment involves collecting and measuring the CO_2 gas formed in the reaction in a 100 cm^3 gas syringe. In order to conduct the experiment safely and to make an accurate measurement, it is necessary to ensure that *no more* than 100 cm^3 of gas is produced in the reaction.

Therefore, it is important to calculate how much sodium hydrogencarbonate should be weighed out at the start, by finding out the mass of sodium hydrogencarbonate required to produce, say, 90 cm^3 of carbon dioxide gas. To do this we assume that the molar volume of carbon dioxide gas is 24 litres mol^{-1}.

$$2NaHCO_3(s) \rightarrow Na_2CO_3(s) + H_2O(g) + CO_2(g)$$

2 mol $\longleftrightarrow$ 1 mol

168 g $\longleftrightarrow$ 24 litres (24 000 cm^3)

$\frac{168 \times 90}{24000}$ $\longleftrightarrow$ 90 cm^3

$= 0.63\,\text{g}$

This calculation shows that the experiment requires approximately 0·63 g of solid sodium hydrogencarbonate at the start.

contd

UNIT 2 PPA 4 – VERIFICATION OF A THERMODYNAMIC PREDICTION contd

Procedure

After weighing out approximately 0·63 g of sodium hydrogen carbonate into a test tube, the apparatus was assembled as shown in the diagram.

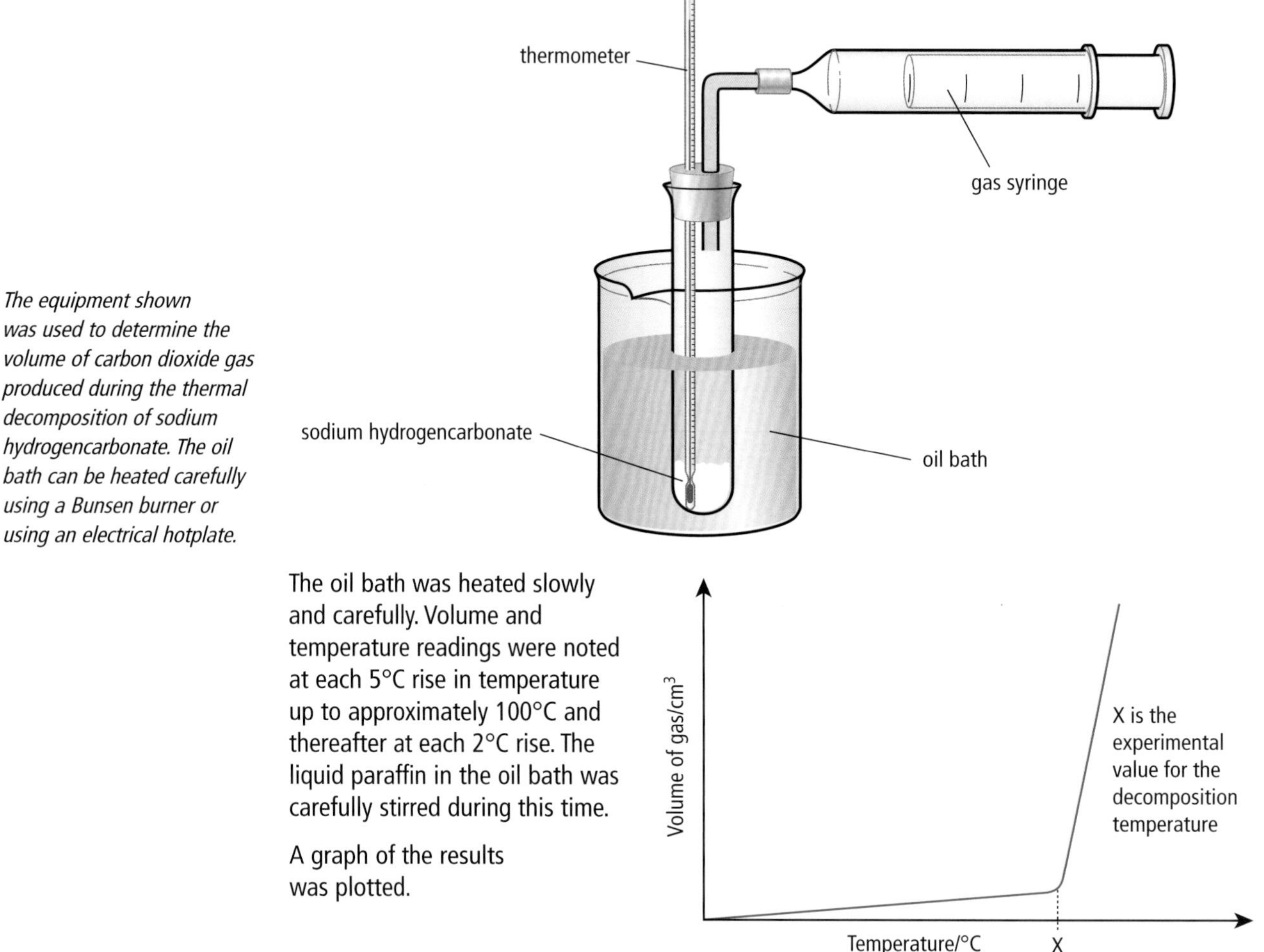

The equipment shown was used to determine the volume of carbon dioxide gas produced during the thermal decomposition of sodium hydrogencarbonate. The oil bath can be heated carefully using a Bunsen burner or using an electrical hotplate.

The oil bath was heated slowly and carefully. Volume and temperature readings were noted at each 5°C rise in temperature up to approximately 100°C and thereafter at each 2°C rise. The liquid paraffin in the oil bath was carefully stirred during this time.

A graph of the results was plotted.

Conclusion

In your conclusion, you would compare your experimental value for the decomposition temperature from your graph with the theoretical value of 112°C. For example, the experimental value of 115°C for the decomposition temperature of sodium hydrogencarbonate compares favourably with the theoretical decomposition temperature of 112°C.

Evaluation

Possible reasons for the difference between the theoretical value and the experimental value for the decomposition temperature include:

- the thermometer not being totally immersed in the solid sodium hydrogencarbonate
- there may be a temperature difference between the thermometer bulb and the solid sodium hydrogen carbonate if the oil was heated too quickly
- the syringe may be sticky due to friction between the glass surfaces
- there may be leaks in the system
- some of the carbon dioxide formed may have dissolved in the condensed water vapour.

DON'T FORGET

You must be able to give reasons for any difference between the theoretical value and the experimental result.

UNIT 2 PPA 5 – KINETICS OF THE ACID-CATALYSED PROPANONE/IODINE REACTION

Aim

To determine the order of reaction with respect to iodine and the value for the rate constant in the acid-catalysed propanone/iodine reaction.

Introduction

The reaction between propanone and iodine,

$$CH_3COCH_3(aq) + I_2(aq) \xrightarrow{H^+(aq)} CH_3COCH_2I(aq) + HI(aq)$$

is first order with respect to propanone and first order with respect to the hydrogen ions which catalyse the reaction. The rate equation is

rate = $k\ [CH_3COCH_3][H^+][I_2]^x$

and the aim of this PPA is to determine the value of x, i.e. the order with respect to iodine.

The experiment started with a reaction mixture in which the initial concentrations of propanone and hydrogen ions were very much larger than that of iodine. This ensured that only the concentration of iodine changed significantly during the reaction. So, we can see what effect this had on the reaction rate.

The course of the reaction was followed by monitoring the concentration of iodine. This was done by removing samples of the reaction mixture and analysing them for iodine by titrating against a standard solution of sodium thiosulphate using starch indicator.

Procedure

50 cm^3 of 0·010 mol l^{-1} iodine solution was pipetted into a clean, dry conical flask which was then stoppered.

25 cm^3 of 1·0 mol l^{-1} sulphuric acid was also pipetted into another clean, dry conical flask.

25 cm^3 of 1·0 mol l^{-1} propanone solution was then pipetted into the conical flask containing the sulphuric acid. This was then stoppered.

Using a measuring cylinder, about 10 cm^3 of approximately 0·5 mol 1^{-1} sodium hydrogencarbonate solution was measured into each of seven different conical flasks.

The sulphuric acid/propanone mixture was added to the iodine solution and the timer started. The flask was stoppered again and swirled to mix the contents.

At fairly regular intervals, 10 cm^3 of the reaction mixture was pipetted into a conical flask containing the sodium hydrogen carbonate solution. The sodium hydrogencarbonate solution is alkaline and neutralises the hydrogen ions which act as a catalyst. As a result, the reaction rate falls dramatically. This is known as **'quenching'**. The time is noted when the pipette is half empty and this is the time elapsed from the start of the reaction.

The 'quenched' solution in the conical flask was titrated against 0·0050 mol l^{-1} sodium thiosulphate, adding starch indicator when the solution became straw-coloured. The end-point was observed when the starch indicator changed from blue-black to colourless.

Further 10 cm^3 samples of the reaction mixture were withdrawn at regular intervals. After quenching and titrating against sodium thiosulphate solution, it was possible to work out the iodine concentrations at the different times. A graph was drawn of iodine concentration against time (see page 60).

A set of experimental results is given in the table on the next page.

DON'T FORGET

The function of the sodium hydrogencarbonate is to neutralise the acid catalyst and so, in effect, to 'stop' the reaction.

The colour change at the end-point was blue–black to colourless.

contd

UNIT 2 PPA 5 – KINETICS OF THE ACID-CATALYSED PROPANONE/IODINE REACTION contd

Time measured/ min	Time/s	Initial burette reading/cm³	Final burette reading/cm³	Vol of $S_2O_3^{2-}$/ cm³
2:41	161	4·0	33·7	29·7
6:47	407	0·0	23·2	23·2
10:57	657	23·2	42·1	18·9
14:44	884	0·1	15·9	15·8
18:40	1120	15·9	28·5	12·6
22:39	1359	28·5	35·0	6·5
26:34	1594	35·0	38·5	3·5

The two ion-electron equations are

$2S_2O_3^{2-} \rightarrow S_4O_6^{2-} + 2e^-$ and $I_2 + 2e^- \rightarrow 2I^-$

giving the overall redox equation

$2S_2O_3^{2-} + I_2 \rightarrow S_4O_6^{2-} + 2I^-$

So 1 mol of I_2 reacts with 2 mol $S_2O_3^{2-}$.

Using the titration results, the following can be calculated:

- $n_{thiosulphate} = V \times c$
- n_{iodine} in 10 cm³ from number of moles of $S_2O_3^{2-}$ used in each titration
- concentration of iodine, $[I_2]$ in mol l⁻¹.

These processed results are shown in the table and graph.

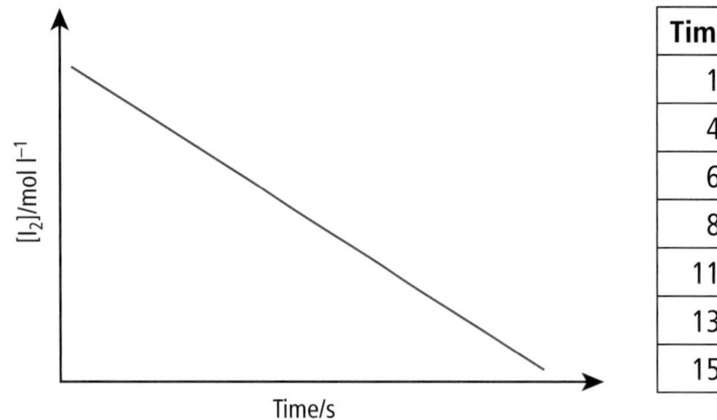

Time/s	$n_{thiosulphate}$	n_{iodine}	$[I_2]$/mol l⁻¹
161	$1{\cdot}48 \times 10^{-4}$	$7{\cdot}4 \times 10^{-5}$	$7{\cdot}4 \times 10^{-3}$
407	$1{\cdot}16 \times 10^{-4}$	$5{\cdot}8 \times 10^{-5}$	$5{\cdot}8 \times 10^{-3}$
657	$9{\cdot}45 \times 10^{-5}$	$4{\cdot}7 \times 10^{-5}$	$4{\cdot}7 \times 10^{-3}$
884	$7{\cdot}9 \times 10^{-5}$	$3{\cdot}95 \times 10^{-5}$	$4{\cdot}0 \times 10^{-3}$
1120	$6{\cdot}3 \times 10^{-5}$	$3{\cdot}15 \times 10^{-5}$	$3{\cdot}2 \times 10^{-3}$
1359	$3{\cdot}25 \times 10^{-5}$	$1{\cdot}63 \times 10^{-5}$	$1{\cdot}6 \times 10^{-3}$
1594	$1{\cdot}75 \times 10^{-5}$	$8{\cdot}75 \times 10^{-6}$	$0{\cdot}88 \times 10^{-3}$

The concentration of iodine plotted against time gave a straight-line graph showing that the reaction was zero order with respect to iodine.

The gradient of the slope can be calculated to give a value for the rate in mol l⁻¹ s⁻¹. Since the gradient of the graph is constant with respect to the iodine concentration, $[I_2]$, we can conclude that the reaction is **zero order** with respect to I_2.

We can now rewrite the rate equation as, rate = $k[CH_3COCH_3][H^+][I_2]^0$ or more simply as rate = $k\,[CH_3COCH_3][H^+]$

This rearranges to give the rate constant,

$$k = \frac{\text{rate}}{[CH_3COCH_3][H^+]}$$

If the gradient of the graph is calculated as $4{\cdot}2 \times 10^{-6}$ mol l⁻¹ s⁻¹ then

$$k = \frac{4{\cdot}2 \times 10^{-6}}{[CH_3COCH_3][H^+]}$$

contd

UNIT 2 PPA 5 – KINETICS OF THE ACID-CATALYSED PROPANONE/IODINE REACTION contd

To calculate the rate constant we need the concentrations of propanone and the hydrogen ions. Initially, these were much higher than the iodine concentration and, therefore, will hardly have changed during the course of the reaction. This means that the initial concentrations of propanone solution and the hydrogen ions can be used to complete the calculation.

At the start of the experiment, the total volume of the reaction mixture was 100 cm^3. This contained 25 cm^3 of 1 mol l^{-1} propanone solution, so the initial concentration of the propanone solution in the reaction mixture was 0·25 mol l^{-1}.

So, $[CH_3COCH_3] = 0{\cdot}25\ mol\ l^{-1}$

The reaction mixture also contained 25 cm^3 of 1·0 mol l^{-1} sulphuric acid. In 1·0 mol l^{-1} sulphuric acid, we assume that the hydrogen ion concentration is 2·0 mol l^{-1}, and so the initial concentration of hydrogen ions in the reaction mixture was 0·5 mol l^{-1}.

So, $[H^+] = 0{\cdot}5\ mol\ l^{-1}$

Substituting these values, $k = \dfrac{4{\cdot}2 \times 10^{-6}}{0{\cdot}25 \times 0{\cdot}5} = 3{\cdot}36 \times 10^{-5}\ l\ mol^{-1}\ s^{-1}$

Conclusion

The reaction is zero order with respect to iodine and the rate constant, k was determined to be $3{\cdot}36 \times 10^{-5}\ l\ mol^{-1}\ s^{-1}$.

Evaluation

There are many sources of possible errors here. For example:

- There would be transfer losses when pouring from one conical flask to the other. This would have an effect on the concentrations of the solutions.
- Noting the time when the pipette is half empty is difficult and so this may produce an error in the measurement of the time.
- The titrations had to be done very accurately as there was no possibility of repeating them to get concordant results and, as in the case of all titrations, it can be difficult to judge the exact end-point.
- The titrations had to be carried out quickly and immediately after the samples had been withdrawn from the reaction mixture. The reason for this is that the reaction between the propanone and iodine continues to take place, albeit very slowly.
- If the reacting solutions used were not accurately prepared or their concentrations were not known accurately, this would affect the final value for the rate constant.
- If the line of best-fit on the graph was not accurately drawn, the gradient would be wrong. The value calculated for the rate would be incorrect as a result and, therefore, the value calculated for the rate constant would also be incorrect.

SYSTEMATIC ORGANIC CHEMISTRY 1 – HYDROCARBONS

SIGMA (σ) AND PI (π) BONDS

A covalent bond is formed when two half-filled atomic orbitals overlap. If they overlap along the axis of the bond (**'end-on'**), a covalent bond known as a **sigma** (σ) bond results.

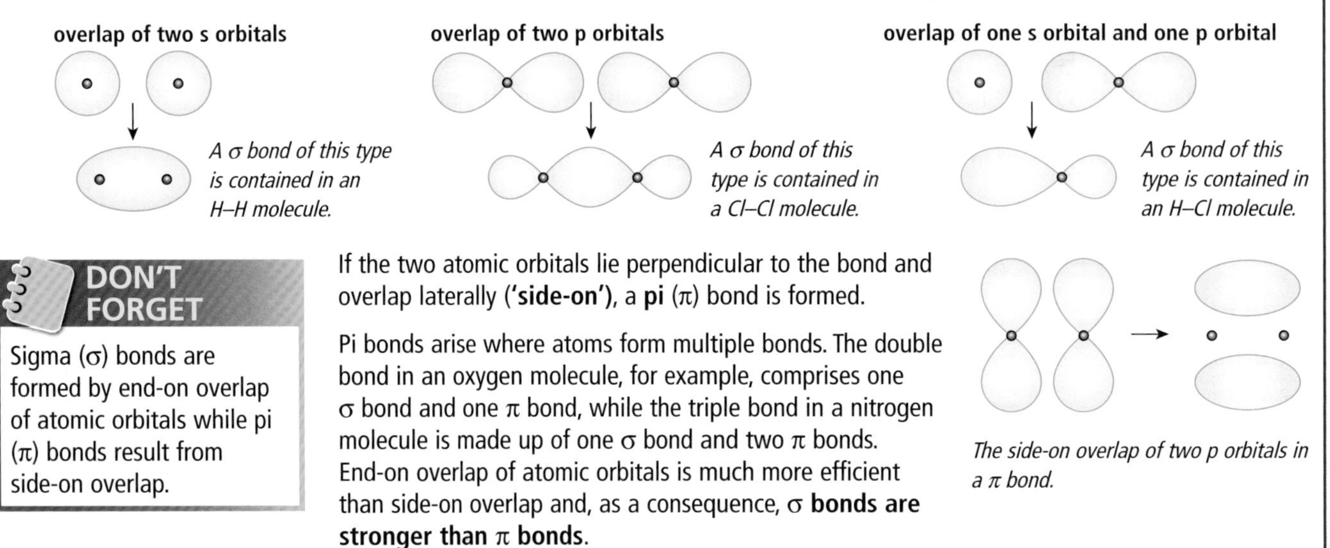

DON'T FORGET

Sigma (σ) bonds are formed by end-on overlap of atomic orbitals while pi (π) bonds result from side-on overlap.

If the two atomic orbitals lie perpendicular to the bond and overlap laterally (**'side-on'**), a **pi** (π) bond is formed.

Pi bonds arise where atoms form multiple bonds. The double bond in an oxygen molecule, for example, comprises one σ bond and one π bond, while the triple bond in a nitrogen molecule is made up of one σ bond and two π bonds. End-on overlap of atomic orbitals is much more efficient than side-on overlap and, as a consequence, **σ bonds are stronger than π bonds**.

The side-on overlap of two p orbitals in a π bond.

HYBRIDISATION

A carbon atom in its ground state has the electronic configuration $1s^2\ 2s^2\ 2p^2$. Since it only has two half-filled orbitals (in the 2p subshell) you might have expected carbon to form two bonds rather than the four we know that it does. To explain the bonding in organic molecules, the theory of hybridisation was introduced. **Hybridisation** is the process of mixing atomic orbitals on an atom to generate a set of new atomic orbitals called **hybrid orbitals**.

To illustrate the theory, consider the bonding in alkanes. Let's take ethane as a typical example. The **three** 2p orbitals and the **one** 2s orbital on each carbon atom mix to form **four** degenerate hybrid orbitals. They are known as **sp^3 hybrid orbitals** and point towards the corners of a tetrahedron in order to minimise repulsion. The four sp^3 hybrid orbitals on each carbon atom overlap end-on with four other orbitals (three hydrogen 1s orbitals and one sp^3 orbital on the other carbon atom) to form four sigma bonds.

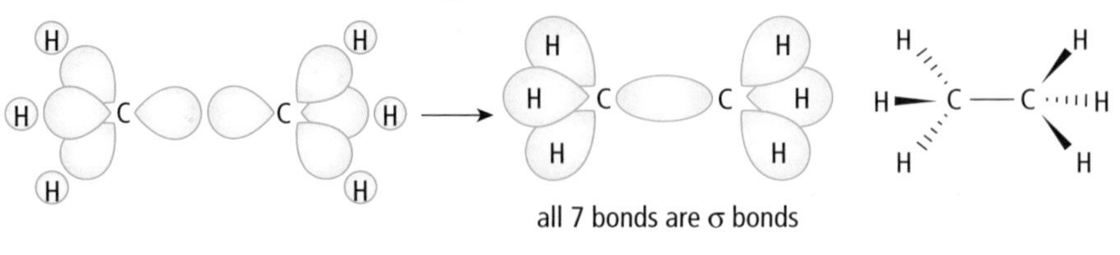

The four sp^3 hybrid orbitals on the carbon atoms overlap end-on with three hydrogen 1s orbitals and one sp^3 orbital on the other carbon atom to give ethane seven sigma bonds.

DON'T FORGET

The bonding in ethane can be described in terms of **sp^3** hybridisation and **sigma** (σ) **bonds** while that in ethene can be described in terms of **sp^2** hybridisation and **sigma** (σ) and **pi** (π) **bonds**.

Consider now the bonding in alkenes and in particular, ethene. The 2s orbital and two of the three 2p orbitals on each carbon atom mix to form three **sp^2 hybrid orbitals**. These hybrid orbitals adopt a trigonal planar arrangement in order to minimise repulsion. Each carbon atom uses its three sp^2 hybrid orbitals to form sigma bonds with two hydrogen atoms and with the other carbon atom. The unhybridised 2p orbitals left on the carbon atoms overlap side-on to form a pi bond.

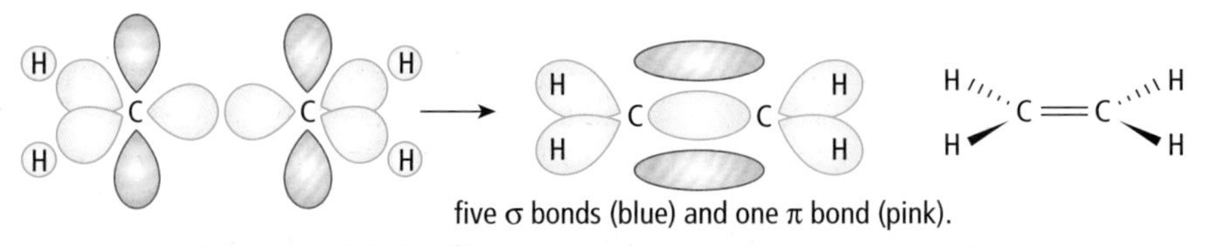

five σ bonds (blue) and one π bond (pink).

ALKANES

Alkanes are not particularly reactive and this is largely due to the non-polar nature of their bonds. They do, however, react with halogens in the presence of sunlight or ultraviolet light. Halogenoalkanes are produced along with 'steamy' fumes of the corresponding hydrogen halide. For example, $CH_4 + Cl_2 \rightarrow CH_3Cl + HCl$

This is an example of a **substitution reaction – a reaction in which an atom or group of atoms in a molecule is replaced by another atom or group of atoms**. In the above reaction, a hydrogen atom has been replaced by a halogen atom. Substitution is not confined to one hydrogen atom per molecule, nor to the same hydrogen atom. As a result, a large number of different halogenoalkanes are produced and so this reaction is of little use in synthetic organic chemistry.

The substitution reaction between an alkane and a halogen is thought to proceed by way of a **chain reaction** which is characterised by three main stages – **initiation**, **propagation** and **termination**.

Initiation step

To illustrate the mechanism, consider the methane–chlorine reaction. Since the reaction does not take place in the dark, ultraviolet light obviously plays a major role in the process. It provides the energy required to break the chlorine–chlorine bond and to split some of the chlorine molecules into atoms.
$Cl{-}Cl \rightarrow Cl\cdot + Cl\cdot$ (A dot (·) represents an unpaired electron)

This type of bond-breaking is known as **homolytic fission**. It tends to occur when there is little or no polarity in the bond. In homolytic fission, one electron of the bond goes to one atom while the other electron goes to the other atom. Chlorine atoms are examples of **radicals** which are **atoms or groups of atoms with an unpaired electron**. Radicals are extremely unstable and therefore highly reactive. So, the initiation step in a chain reaction is the process whereby radicals are generated.

Propagation steps

Each chlorine radical that is produced goes on to attack a methane molecule, removing a hydrogen atom from it to form hydrogen chloride and a methyl radical. The latter then attacks a chlorine molecule to form chloromethane and another chlorine radical:
(i) $Cl\cdot + CH_4 \rightarrow HCl + CH_3\cdot$ (ii) $CH_3\cdot + Cl_2 \rightarrow CH_3Cl + Cl\cdot$

In each of the propagation steps, one radical enters the reaction and another radical is generated. These steps 'propagate' or sustain the chain reaction.

Termination steps

As the number of radicals builds up, collisions between them occur and stable molecules are produced.
(i) $CH_3\cdot + Cl\cdot \rightarrow CH_3Cl$ (ii) $CH_3\cdot + CH_3\cdot \rightarrow CH_3CH_3$ (iii) $Cl\cdot + Cl\cdot \rightarrow Cl_2$

The termination steps are reactions in which radicals are used up and not regenerated. Such reactions bring the chain reaction to an end.

DON'T FORGET

A chain reaction is characterised by the following steps:
- initiation
- propagation
- termination

LET'S THINK ABOUT THIS

In a recent examination, this question was asked.
Which of the following is a propagation step in the chlorination of methane?

A $Cl_2 \rightarrow Cl\cdot + Cl\cdot$
B $CH_3\cdot + Cl\cdot \rightarrow CH_3Cl$
C $CH_3\cdot + Cl_2 \rightarrow CH_3Cl + Cl\cdot$
D $CH_4 + Cl\cdot \rightarrow CH_3Cl + H\cdot$

A propagation step in a chain reaction involves a radical as a reactant and a radical as a product. So, A (the initiation step) and B (a termination step) can be eliminated. If you recall the detailed mechanism (see above) you'll arrive at C as the answer.

D, however, fits the definition, so why is it wrong? Let's consider the feasibility of this reaction. $\Delta S^{\ominus}$ will be approximately zero since there is no change in the number of moles of gas in the transition from reactants to products. $\Delta G^{\ominus}$ will, therefore, be approximately equal to $\Delta H^{\ominus}$. The reaction involves breaking 1 mole of C–H bonds (+414 kJ) and making 1 mole of C–Cl bonds (–326 kJ) giving a $\Delta H^{\ominus}$ value and hence a $\Delta G^{\ominus}$ value of +88 kJ mol^{-1}. Since $\Delta G^{\ominus}$ is positive, the reaction is not feasible. Check out the $\Delta G^{\ominus}$ value for the reaction shown in C; you should get –83 kJ mol^{-1}.

SYSTEMATIC ORGANIC CHEMISTRY 2 – ALKENES

PREPARATION OF ALKENES

Alkenes can be prepared by

- dehydration of alcohols
- base-induced elimination of hydrogen halides from monohalogenoalkanes.

Dehydration of alcohols

The vapour of the alcohol can be passed over hot aluminium oxide, or the alcohol can be treated with concentrated sulphuric acid or concentrated phosphoric acid (orthophosphoric acid).

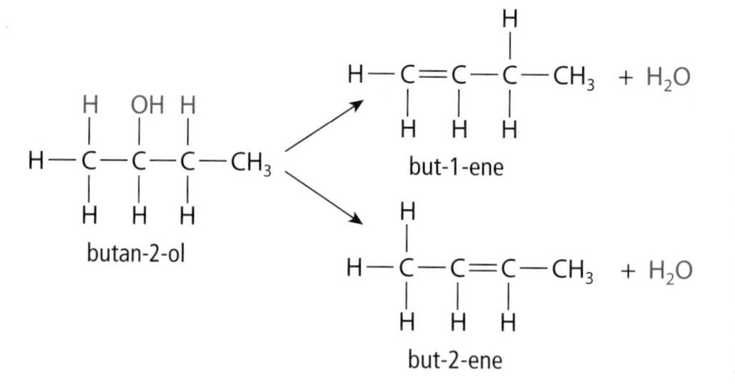

During dehydration, the –OH group is removed, along with an H atom on an adjacent carbon atom. In this case, two alkenes are formed, although but-2-ene is the major product. With some alcohols, such as butan-1-ol and propan-2-ol, only one alkene is produced.

This reaction can also be described as an **elimination reaction**, i.e. a reaction in which the elements of a simple molecule, like water, are removed from the organic molecule and not replaced.

Concentrated phosphoric acid is preferred to concentrated sulphuric acid when dehydrating alcohols, since with the latter more side reactions occur and it tends to lead to charring.

Base-induced elimination of hydrogen halides from monohalogenoalkanes

This is achieved by heating the monohalogenoalkane, under reflux, with ethanolic potassium (or sodium) hydroxide (a solution of potassium (or sodium) hydroxide in ethanol).

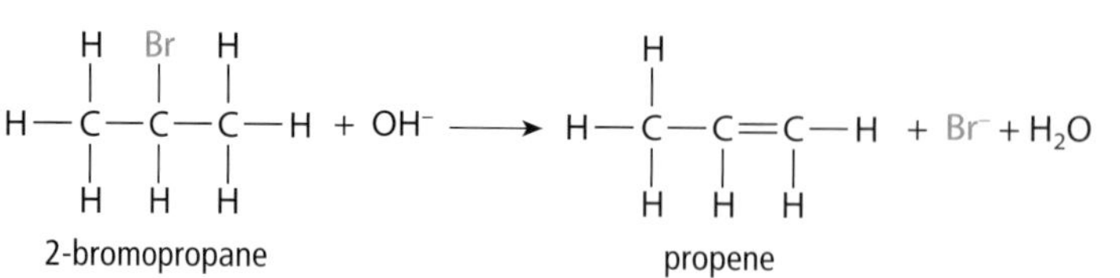

The reaction is referred to as a base-induced elimination since it is the presence of the base, potassium hydroxide, which drives the reaction.

Like the dehydration of some alcohols, the elimination of hydrogen halides from monohalogenoalkanes can result in the formation of two alkenes. For example, heating 2-chlorobutane with ethanolic potassium hydroxide produces but-1-ene and but-2-ene (major product).

In the elimination of a hydrogen halide from a monohalogenoalkane, the base, potassium hydroxide or sodium hydroxide, must be dissolved in ethanol and **not** water.

REACTIONS OF ALKENES

Alkenes undergo **addition reactions** with **hydrogen** to form **alkanes**, with **halogens** to form **dihaloalkanes**, with **hydrogen halides** to form **monohalogenoalkanes** and with **water** to form **alcohols**.

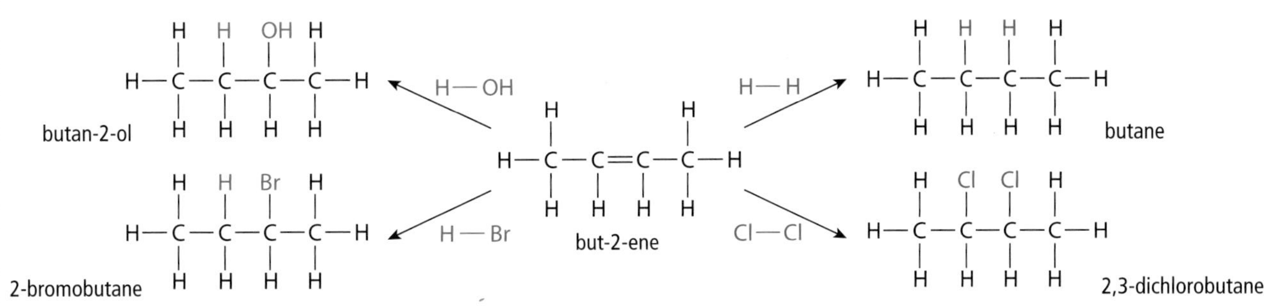

The addition reaction of an **alkene** with

- **hydrogen** is also known as **hydrogenation**. It is catalysed by nickel or platinum.
- a **halogen** is also known as **halogenation**.
- a **hydrogen halide** is also known as **hydrohalogenation**.
- **water** is also known as **hydration**. It is catalysed by acid.

Alkenes can undergo hydrogenation, halogenation, hydrohalogenation and hydration.

MARKOVNIKOV'S RULE

When H–H, Br–Br, H–Br or H–OH is added to but-2-ene only one product is formed (see above). However, when a hydrogen halide or water is added to an unsymmetrical alkene, i.e. one in which the groups of atoms attached to one carbon atom of the double bond are not identical to the groups attached to the other carbon atom, two products are formed. For example, when hydrogen chloride is added to the unsymmetrical alkene, but-1-ene, both 2-chlorobutane and 1-chlorobutane are formed.

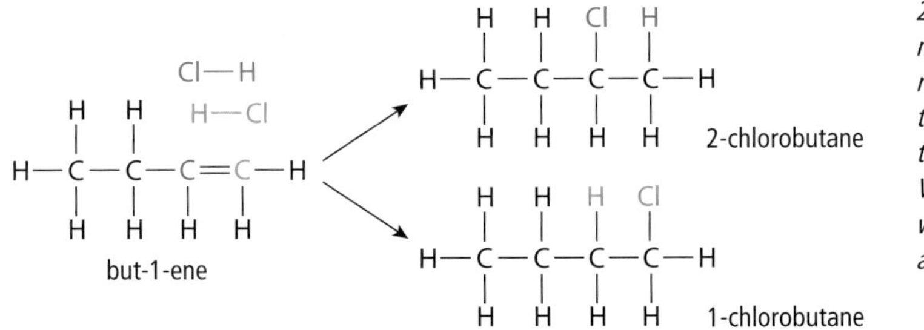

2-Chlorobutane is the major product of this reaction, but how can this be predicted? It was the Russian chemist, Vladimir Markovnikov, who formulated a rule to answer this question.

Markovnikov's rule states that when H–X or H–OH adds onto an unsymmetrical alkene, the major product is the one in which the H atom ends up attached to the carbon atom of the double bond that already has the greater number of hydrogen atoms bonded to it. If you look back at the structure of but-1-ene, the carbon atom (coloured green) on the right of the double bond has more H atoms attached than the one on the left (coloured red). So, on adding hydrogen chloride, the H atom of the HCl molecule will attach itself to the right-hand carbon atom of the double bond giving 2-chlorobutane as the major product. Markovnikov's rule is empirical and only when the mechanisms of these addition reactions were worked out was an explanation forthcoming (see page 67).

DON'T FORGET

Markovnikov's rule states that the main product of the reaction between an unsymmetrical alkene and water, or an unsymmetrical alkene and a hydrogen halide, is the one in which the hydrogen atom adds to the carbon atom of the double bond with the larger number of hydrogen atoms already attached.

LET'S THINK ABOUT THIS

1-Methylcyclohexene reacts with hydrogen chloride to form two products. Draw a structural formula for each of the two products and name the major product.

SYSTEMATIC ORGANIC CHEMISTRY 3 – ADDITION REACTION MECHANISMS

HALOGENATION

The mechanism for the halogen–alkene addition reaction is a two-step process and can be illustrated using bromine and ethene.

Step 1

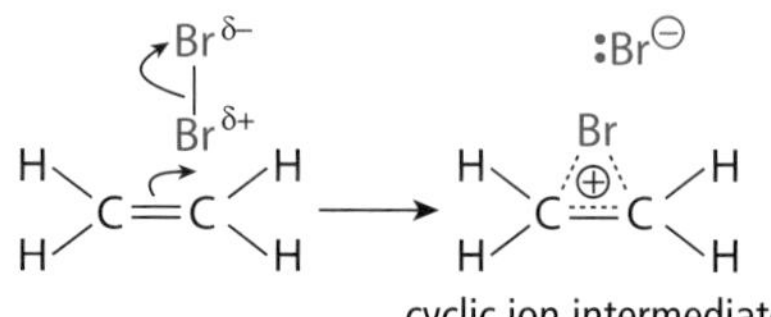

The first step in the reaction of bromine and ethene results in the formation of a cyclic ion intermediate.

As the bromine molecule approaches the double bond in ethene, it becomes polarised. The electron-rich double bond in ethene 'pushes' the electrons in the bromine molecule towards the Br atom which is remote from the double bond (see diagram). This Br atom gains a slight negative charge, leaving the other Br atom with a slight positive charge. The $Br^{\delta+}$ atom of the bromine molecule attacks the ethene and a **cyclic ion intermediate** is formed, along with a bromide ion.

In this step, the bromine molecule is described as an **electrophile (an electron-pair acceptor)**. Electrophilic means 'electron loving' and so an electrophile will seek out a centre of negative charge such as the double bond in ethene.

The bond in the bromine molecule breaks and it is said to have undergone **heterolytic fission** since one atom of the bond retains both bonding electrons. As a result, ions are formed.

The curly arrows (↷) are used to indicate the movement of electron pairs in a reaction. The tail of the arrow shows the origin of the electrons that move and the head shows where they end up.

DON'T FORGET

Nucleophiles are molecules or negatively charged ions which have at least one lone pair of electrons that they can donate. Electrophiles are molecules or positively charged ions which are capable of accepting an electron pair.

Step 2

The Br^- ion attacks the cyclic intermediate ion from the side opposite to where the Br atom is attached. This happens because the bromine atom in the intermediate is large and so prevents access to that side of the cyclic ion intermediate. In this second step, the Br^- ion is acting as a **nucleophile (an electron-pair donor)**. A nucleophile seeks out a centre of positive charge in another species.

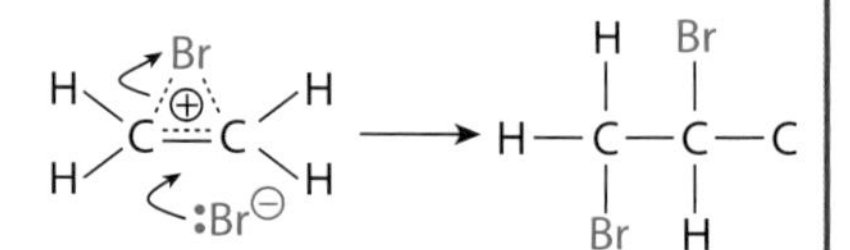

The second step of the mechanism involves the Br^- ion attacking the cyclic ion intermediate.

HYDROHALOGENATION

The mechanism for the hydrogen halide–alkene addition reaction is again a two-step process and is illustrated below using hydrogen bromide and propene.

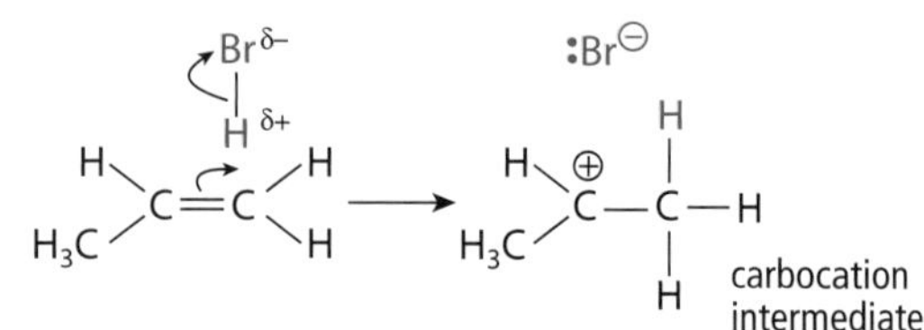

Step 1 in the hydrohalogenation reaction is the formation of a carbocation.

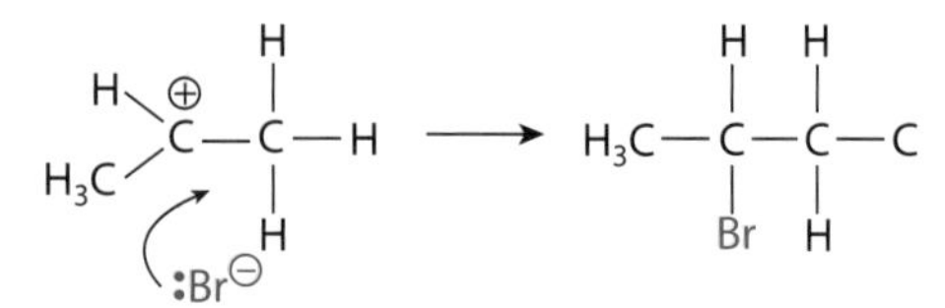

Step 2 involves the Br^- ion attacking the carbocation intermediate.

Step 1

In this case, the H–Br molecule is already polarised. Its $H^{\delta+}$ atom attacks the double bond in propene and bonds to one of the carbon atoms of the double bond forming an intermediate known as a **carbocation**. At the same time, the bond in the H–Br molecule breaks heterolytically and a bromide ion is formed as well.

Step 2

The second step involves the Br^- ion attacking the carbocation intermediate. In this mechanism, unlike the halogenation reaction shown above, the Br^- ion can attack from either side of the carbocation.

The product here is 2-bromopropane, but you'll remember that when a halogen reacts with an unsymmetrical alkene, like propene, two products are formed. The other product is I-bromopropane.

contd

HYDROHALOGENATION contd

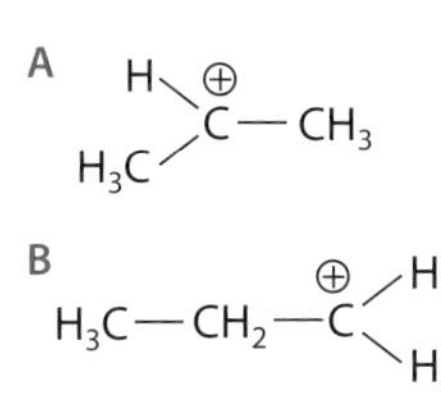

The two carbocations that would be formed are shown on the right. The stability of a carbocation depends on the number of alkyl groups attached to the positively charged carbon atom. Alkyl groups are electron-donating and can push electrons onto the positively charged carbon atom, thus increasing the stability of the carbocation. So the more alkyl groups attached to the positively charged carbon atom, the greater the stability of the carbocation. Carbocation A is therefore more stable than carbocation B and, consequently, 2-bromopropane, rather than 1-bromopropane, will be the major product of the reaction. This helps to explain why Markovnikov's rule applies to the addition of a hydrogen halide to an unsymmetrical alkene.

ACID-CATALYSED HYDRATION

The mechanism for the acid-catalysed hydration reaction is very similar to that for the hydrohalogenation of alkenes and also proceeds via a carbocation intermediate. It is outlined here using water and propene.

Step 1

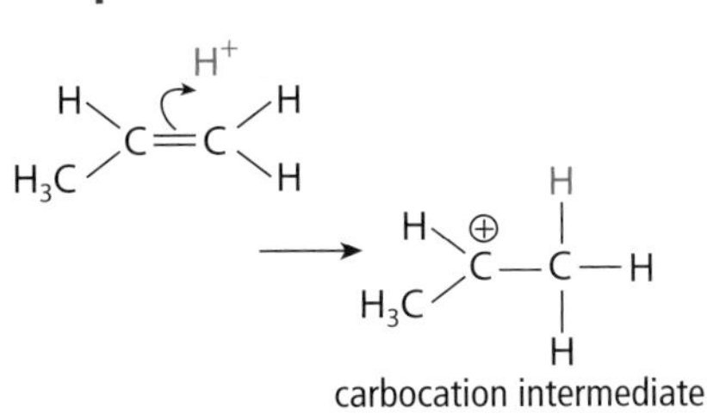

The hydrogen ion of the acid catalyst is an electrophile and attacks the electron-rich double bond in the propene molecule to form a carbocation.

Step 2

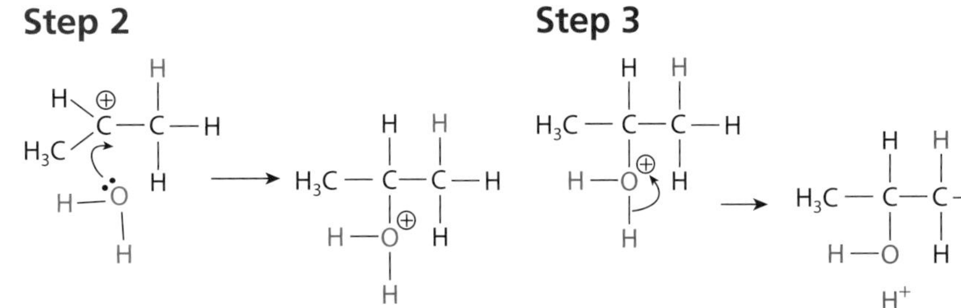

The carbocation then undergoes rapid nucleophilic attack by a water molecule to give a protonated alcohol (an alcohol with a hydrogen ion attached).

Step 3

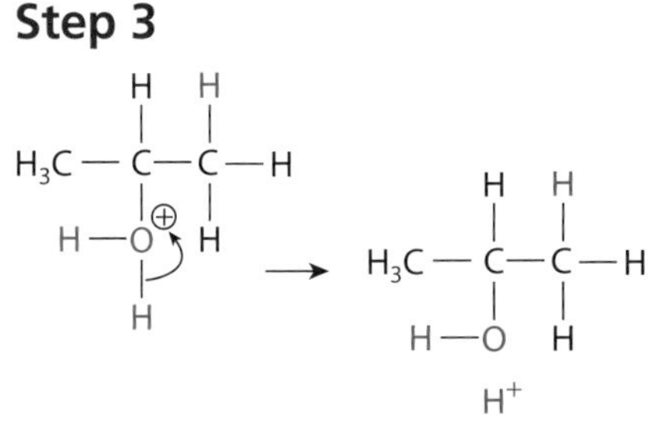

The protonated propan-2-ol is a strong acid and readily loses a proton (hydrogen ion) to give the final product, propan-2-ol.

You will notice that hydrogen ions are used up in the first step and regenerated in the final step, thus confirming their role as catalysts.

Like the hydrohalogenation reaction, the hydration of alkenes leads to two products. As well as propan-2-ol, propan-1-ol is also formed. In accordance with Markovnikov's rule, propan-2-ol is the major product. This is again due to the greater stability of carbocation A compared with carbocation B.

A: $(H)(H_3C)C^{\oplus}-CH_3$ B: $H_3C-CH_2-C^{\oplus}H_2$

Carbocation A is more stable than carbocation B.

DON'T FORGET

The mechanism for the halogenation of an alkene proceeds via a cyclic ion intermediate while the mechanisms for the hydrohalogenation and acid-catalysed hydration of alkenes proceed via carbocation intermediates.

LET'S THINK ABOUT THIS

In the early days of alkene chemistry, some researchers found that the hydrohalogenation of alkenes followed Markovnikov's rule, while others found that the same reaction did not. For example, when freshly distilled but-1-ene was exposed to hydrogen bromide, the major product was 2-bromopropane as expected by the Markovnikov rule. However, when the same reaction was carried out with a sample of but-1-ene which had been exposed to air, the major product was 1-bromopropane formed by anti-Markovnikov addition. This caused considerable confusion, but the mystery was solved by the American chemist, Morris Kharasch, in the 1930s. He realised that the samples of alkenes which had been stored in the presence of air had formed peroxide radicals and that the hydrohalogenation proceeded by a radical chain reaction mechanism and not via the mechanism involving carbocation intermediates as when pure alkenes were used.

http://www.youtube.com/watch?v=EWOvFAu8FmA Markovnikov tutorial

SYSTEMATIC ORGANIC CHEMISTRY 4 – HALOGENOALKANES

NAMING AND STRUCTURAL TYPES

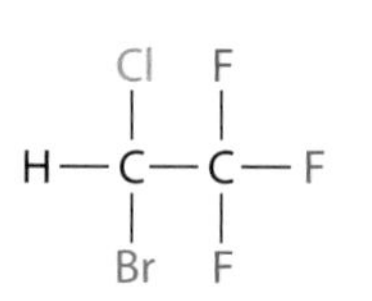

2-bromo-2-chloro-1,1,1-trifluoroethane or halothane, an anaesthetic.

Halogenoalkanes can be regarded as substituted alkanes in which one or more of the hydrogen atoms in the alkane are replaced by a halogen atom. They are named in a similar fashion to branched-chain alkanes with the halogen atoms treated like branches. For example, the anaesthetic halothane has the structure shown in the diagram and is called 2-bromo-2-chloro-1,1,1-trifluoroethane.

There are three structural types of monohalogenoalkanes – primary, secondary and tertiary. They are classified according to the number of alkyl groups attached to the carbon atom bearing the halogen atom.

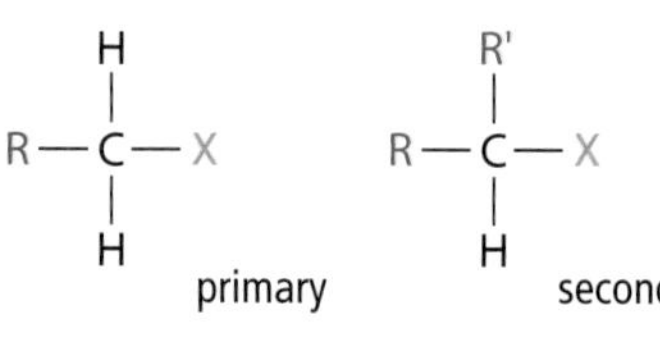

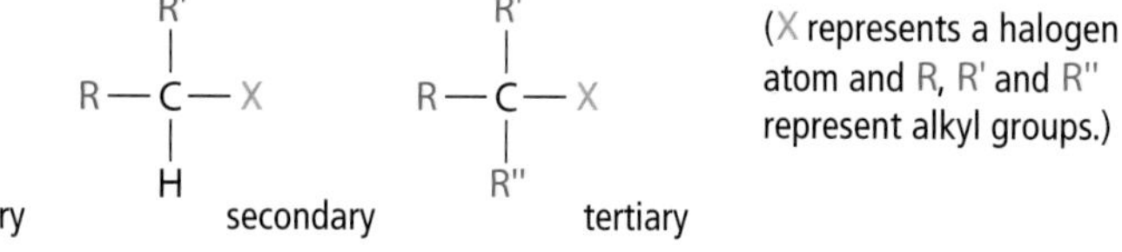

(X represents a halogen atom and R, R' and R'' represent alkyl groups.)

The primary monohalogenoalkane has one alkyl group attached to the halogen-bearing carbon atom, the secondary has two and the tertiary has three.

REACTIONS OF HALOGENOALKANES

δ+ δ–
—C—X

The polar nature of the carbon–halogen bond makes halogenalkanes susceptible to nucleophilic attack.

The main type of reaction which halogenoalkanes undergo is **nucleophilic substitution**.

This is due to the polar nature of the carbon–halogen bond. It is the presence of the slight positive charge on the carbon atom of the C–X bond that makes halogenoalkanes susceptible to attack by **nucleophiles**.

The nucleophile donates an electron pair and, in so doing, forms a covalent bond with the carbon atom of the C–X bond. At the same time, the halogen atom is 'thrown out' and **substituted** by the nucleophile.

Some nucleophilic substitution reactions of monohalogenoalkanes

The examples shown here use 1-chloropropane as a typical halogenoalkane. The nucleophiles are shown in red and the reagents in blue.

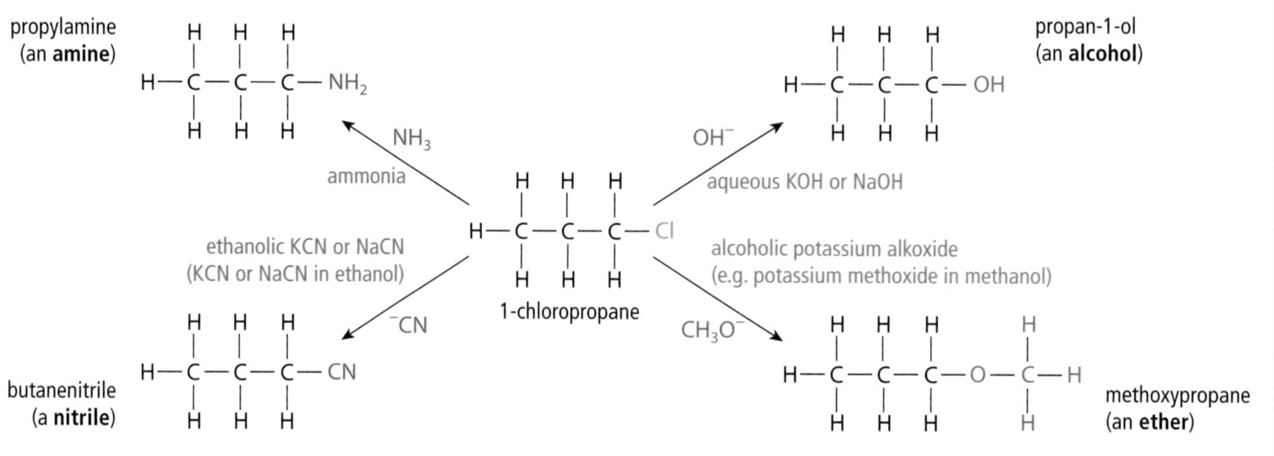

Alkoxides are formed when an alkali metal, such as sodium and potassium, is added to an alcohol. For example, when potassium is added to methanol, potassium methoxide is formed.

$$2K + 2CH_3OH \rightarrow H_2 + 2CH_3O^-K^+$$

contd

REACTIONS OF HALOGENOALKANES contd

You'll have noticed that the **nitrile** formed when 1-chloropropane is treated with ethanolic potassium (or sodium) cyanide contains one more carbon atom than the original halogenoalkane. This makes the reaction very useful in synthetic organic chemistry, since it is a means of increasing the chain length of an organic compound. The nitrile produced in the reaction can be converted into the corresponding carboxylic acid by acid hydrolysis.

$$CH_3CH_2CH_2CN \xrightarrow{H_2O/H^+} CH_3CH_2CH_2COOH$$

butanenitrile → butanoic acid

A nitrile can be converted into a carboxylic acid by acid hydrolysis.

Nucleophilic substitution mechanisms

A halogenoalkane will undergo nucleophilic substitution by one or other of two different mechanisms labelled $\mathbf{S_N1}$ and $\mathbf{S_N2}$.

To illustrate the $\mathbf{S_N1}$ **mechanism**, consider the reaction between the tertiary halogenoalkane, 2-bromo-2-methylpropane, and the nucleophilic hydroxide ion. A study of the kinetics of the reaction shows that it has the following rate equation: rate = $k[(CH_3)_3CBr]$

It is first order with respect to the halogenoalkane and this implies that the rate-determining step can **only** involve the halogenoalkane.

The S_N1 mechanism is a two-step process.

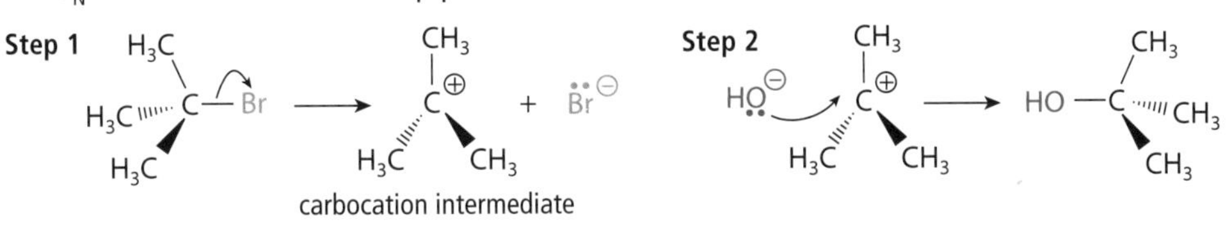

Step 1 is the slow rate-determining step as it involves only the halogenoalkane.

> **DON'T FORGET**
> Monohalogenoalkanes react with alkalis to form alcohols, with alcoholic alkoxides to form ethers, with ethanolic cyanides to form nitriles and with ammonia to form amines.

To illustrate the $\mathbf{S_N2}$ **mechanism**, consider the reaction between the primary halogenoalkane, bromoethane, and the nucleophilic hydroxide ion. A study of the kinetics of the reaction shows that it has the following rate equation: rate = $k[CH_3CH_2Br][OH^-]$

It is first order with respect to the halogenoalkane and first order with respect to the hydroxide ion. This implies that the rate-determining step involves both the halogenoalkane and the hydroxide ion.

The S_N2 mechanism involves only one step.

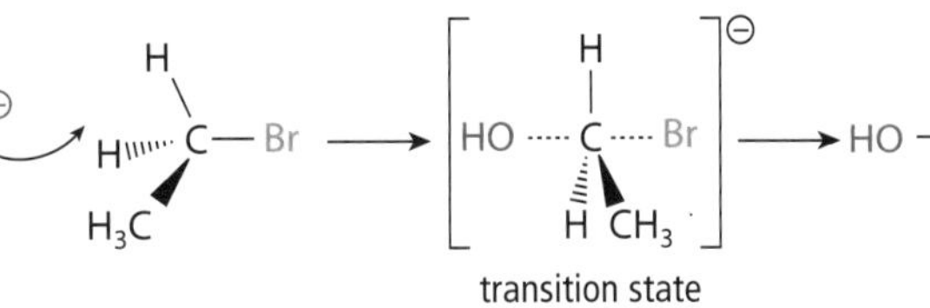

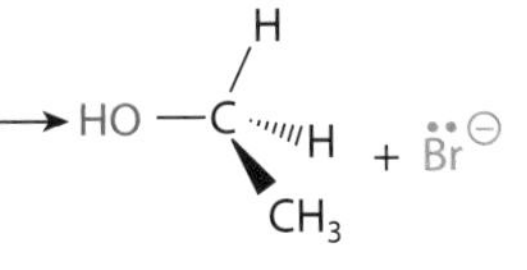

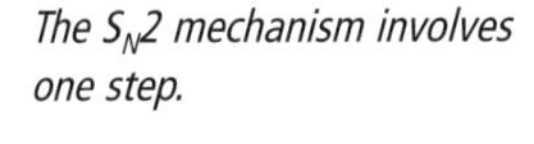

The S_N2 mechanism involves one step.

When a halogenoalkane undergoes nucleophilic substitution, how can we tell whether it will proceed via an S_N1 or S_N2 mechanism? One factor which will help us decide is the structural type of the halogenoalkane; whether it is a primary, secondary or tertiary halogenalkane.

In the S_N1 mechanism a carbocation intermediate is formed which could be a primary, secondary or tertiary carbocation. Since alkyl groups are electron-donating, the tertiary carbocation will be the most stable – so tertiary halogenoalkanes are most likely to react via an S_N1 mechanism. Primary halogenoalkanes are least likely to react via this mechanism.

In the S_N2 mechanism, the nucleophile attacks the carbon atom of the C–X bond from the side opposite to the halogen atom (see above). In the case of a tertiary halogenoalkane, attack from that side is likely to be hindered by three bulky alkyl groups. So, tertiary halogenolkanes are least likely to react via an S_N2 mechanism and primary halogenoalkanes most likely.

> **DON'T FORGET**
> $\mathbf{S_N1}$ means **S**ubstitution, **N**ucleophilic, **1st** order while $\mathbf{S_N2}$ means **S**ubstitution, **N**ucleophilic, **2nd** order.

LET'S THINK ABOUT THIS

Suggest why the compound drawn is unlikely to react with aqueous sodium hydroxide via an S_N2 mechanism.

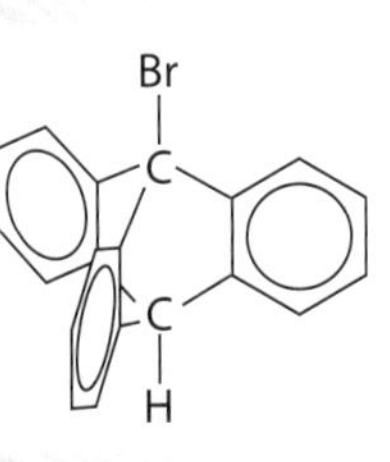

If the mechanism was S_N2, then the OH^- ion nucleophile would have to attack the carbon atom of the C–Br bond from the side exactly **opposite** to that of the Br atom. You can see from the structure that access to this C atom would be severely restricted, thus explaining why the compound is unlikely to react via an S_N2 mechanism.

SYSTEMATIC ORGANIC CHEMISTRY 5 – ALCOHOLS AND ETHERS

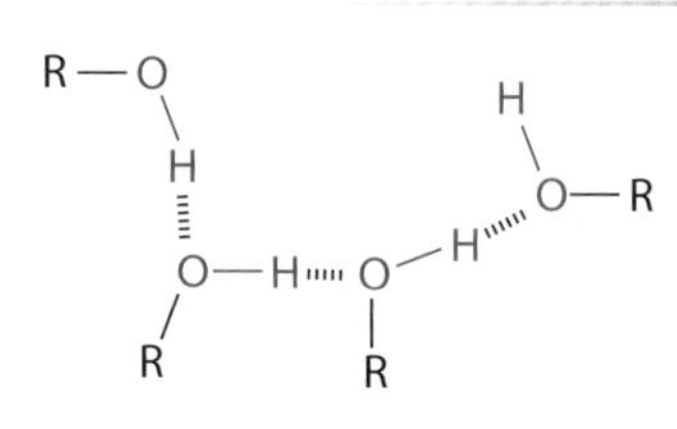

''''''''' represents hydrogen bonding

Hydrogen bonds between alcohol molecules explain the higher boiling point of alcohols compared to alkanes of similar molecular mass and shape.

PHYSICAL PROPERTIES OF ALCOHOLS

As with alkanes, the **boiling points of alcohols** show a progressive increase with the addition of each $-CH_2-$ unit, but alcohols have considerably higher boiling points than alkanes of similar relative formula mass and shape. The reason is the presence of the **polar –OH group** in the alcohol molecule which allows **hydrogen bonds** to be set up between the molecules as shown in the diagram.

Since hydrogen bonds are stronger than van der Waals' forces, extra energy is needed to break them, accounting for the higher boiling points of the alcohols.

There is also a gradation in the **solubilities of alcohols** in water. The lower alcohols (methanol, ethanol and propan-1-ol) are completely soluble in water (miscible with water), but heptan-1-ol and other higher alcohols are insoluble in water. The lower alcohols dissolve in water because the energy released when forming hydrogen bonds between the alcohol and water molecules is sufficient to break the hydrogen bonds between the water molecules. However, the increasing length of the alkyl group means that by the time heptan-1-ol is reached, this is no longer true since the large non-polar hydrocarbon part of the molecule masks the polar hydroxyl group.

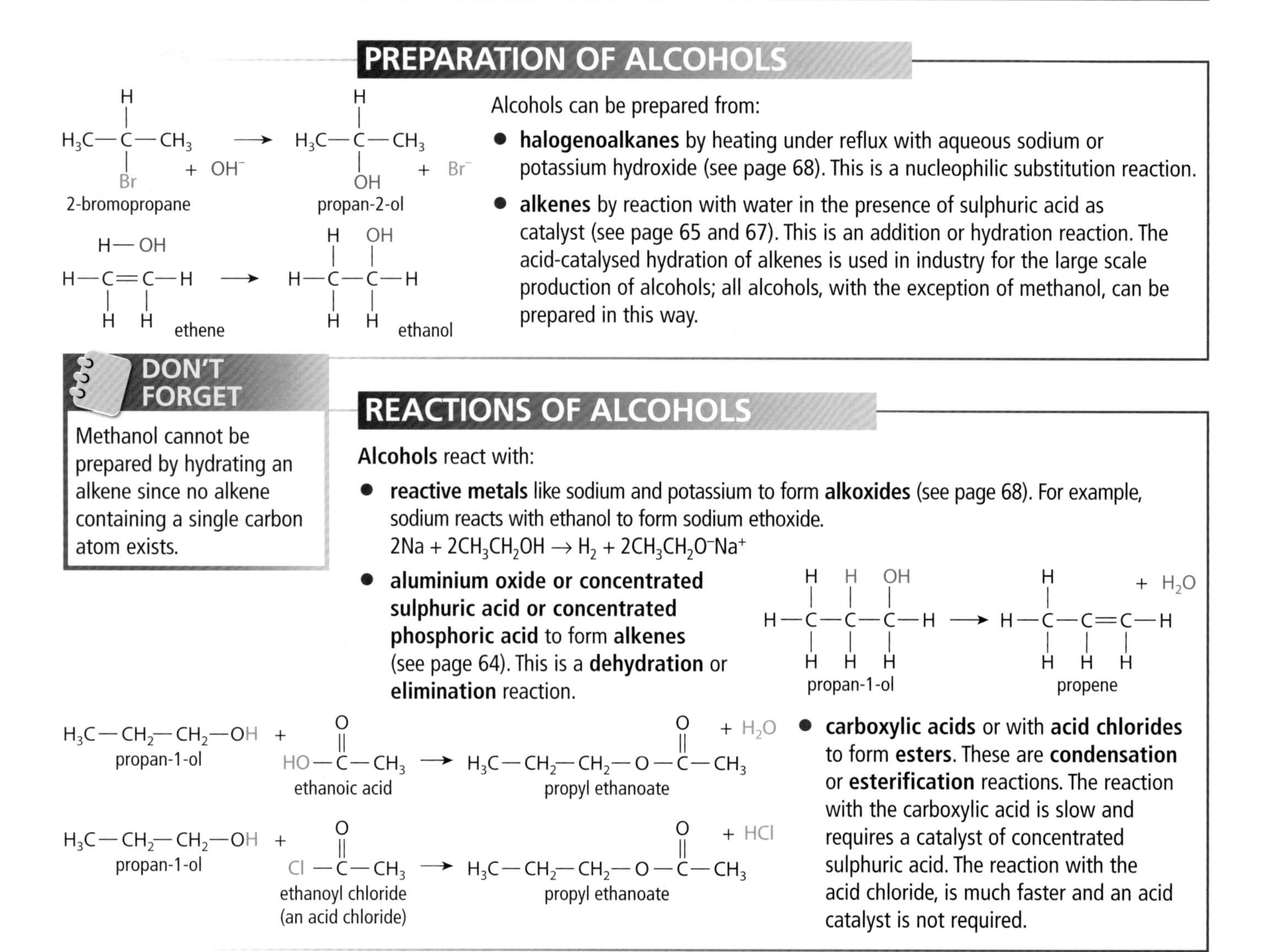

PREPARATION OF ALCOHOLS

Alcohols can be prepared from:

- **halogenoalkanes** by heating under reflux with aqueous sodium or potassium hydroxide (see page 68). This is a nucleophilic substitution reaction.
- **alkenes** by reaction with water in the presence of sulphuric acid as catalyst (see page 65 and 67). This is an addition or hydration reaction. The acid-catalysed hydration of alkenes is used in industry for the large scale production of alcohols; all alcohols, with the exception of methanol, can be prepared in this way.

DON'T FORGET

Methanol cannot be prepared by hydrating an alkene since no alkene containing a single carbon atom exists.

REACTIONS OF ALCOHOLS

Alcohols react with:

- **reactive metals** like sodium and potassium to form **alkoxides** (see page 68). For example, sodium reacts with ethanol to form sodium ethoxide.
  $2Na + 2CH_3CH_2OH \rightarrow H_2 + 2CH_3CH_2O^-Na^+$
- **aluminium oxide or concentrated sulphuric acid or concentrated phosphoric acid** to form **alkenes** (see page 64). This is a **dehydration** or **elimination** reaction.
- **carboxylic acids** or with **acid chlorides** to form **esters**. These are **condensation** or **esterification** reactions. The reaction with the carboxylic acid is slow and requires a catalyst of concentrated sulphuric acid. The reaction with the acid chloride, is much faster and an acid catalyst is not required.

NAMING ETHERS

Ethers have the general formula **R–O–R′** where **R** and **R′** are alkyl groups. When R and R′ are different, the ether is described as unsymmetrical; when they are identical the ether is said to be symmetrical. An ether can be regarded as an alkane with an alkoxy group attached and is named as such.

Alkoxy groups are named by removing the 'yl' from the name of the alkyl substituent and adding 'oxy'. So, CH_3O– is named methoxy and CH_3CH_2O– is named ethoxy. To name the ether, the longest carbon chain is identified to give the parent name. This is prefixed by the name of the alkoxy substituent. Consider, for example, the ether with the structure drawn on the right.

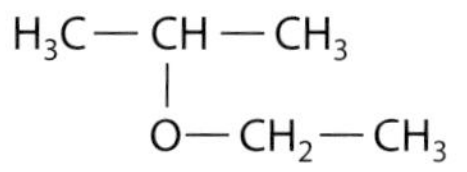

$H_3C—CH(—O—CH_2—CH_3)—CH_3$

2-ethoxypropane

The longest chain contains three carbon atoms and so the parent name is propane. The ethoxy group is attached to the second carbon atom, so this ether is called 2-ethoxypropane. The number '2' is used to distinguish it from ethoxypropane which has the structure here.

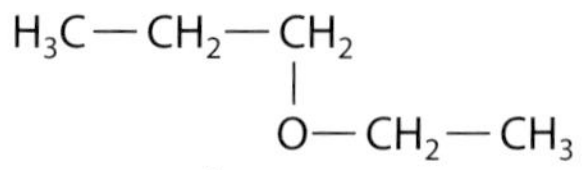

$H_3C—CH_2—CH_2—O—CH_2—CH_3$

ethoxypropane

PHYSICAL PROPERTIES OF ETHERS

Ethers have much lower boiling points than their isomeric alcohols because hydrogen bonding does not occur between ether molecules. The reason for this is that the highly electronegative oxygen atom is not directly bonded to a hydrogen atom. Ether molecules, however, can form hydrogen bonds with water molecules as shown in the diagram.

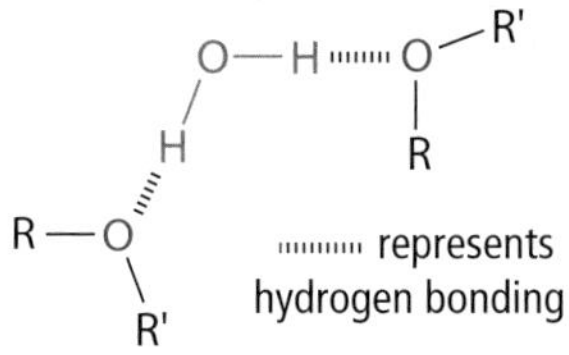

R—O(R')⋯H—O—H⋯O(R)(R')

⋯ represents hydrogen bonding

Hydrogen bonding can occur between water and ether molecules, but not between ether molecules themselves.

Hydrogen bonding with water molecules explains why those ethers with low relative formula masses, like methoxymethane and methoxyethane, are soluble in water. The larger ethers are insoluble in water – making them useful in extracting organic compounds from aqueous solutions.

Ethers are volatile, highly flammable and when they are exposed to air they slowly form peroxides which are unstable and can be explosive. Ethers are used as solvents since they are relatively inert chemically and most organic compounds dissolve in them. Being volatile, these solvents are easily removed by distillation.

Ether molecules cannot form hydrogen bonds with other ether molecules, but they can form hydrogen bonds with water molecules.

PREPARATION OF ETHERS

Ethers can be prepared by refluxing a halogenoalkane with an alkoxide (see page 68). For example, ethoxyethane is made by heating chloroethane with a solution of potassium ethoxide in ethanol. The reaction taking place is nucleophilic substitution.

$$\underset{\text{chloroethane}}{H_3C—CH_2—Cl} + {}^{-}O—CH_2—CH_3 \longrightarrow \underset{\text{ethoxyethane}}{H_3C—CH_2—O—CH_2—CH_3} + Cl^{-}$$

LET'S THINK ABOUT THIS

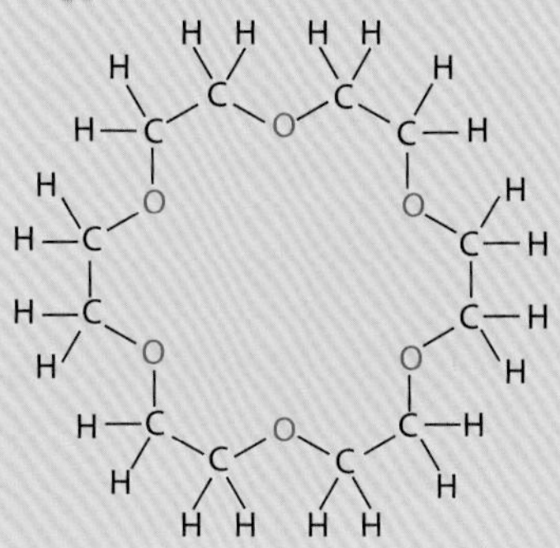

In recent years, there has been a lot of interest in cyclic ethers like the one shown here.

Known as crown ethers because of their crown-like shape, these ethers contain cavities that are ideal for complex formation with metal ions. It is this property which allows ordinary salts to dissolve in organic solvents. For example, potassium permanganate is usually insoluble in benzene but readily dissolves in it if [18]-crown-6 ether is added. This solution is useful because it allows oxidation with potassium permanganate to be carried out in organic solvents. The potassium ion (shown in green in this diagram) is just the right size to fit into cavity in the crown ether.

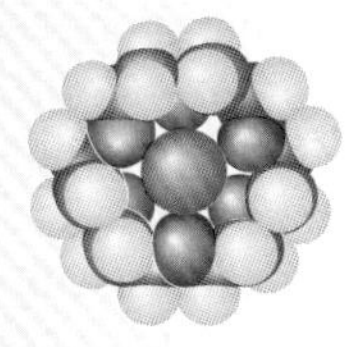

SYSTEMATIC ORGANIC CHEMISTRY 6 – ALDEHYDES AND KETONES

STRUCTURES AND PHYSICAL PROPERTIES OF ALDEHYDES AND KETONES

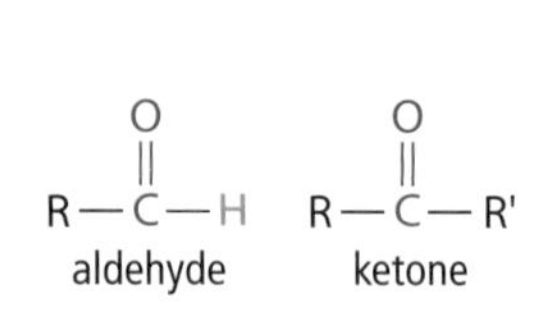

Aldehydes and ketones share the **carbonyl functional group**. In aldehydes a hydrogen atom is attached to the carbonyl group, whereas in ketones the carbonyl group is flanked by two carbon atoms.

Aldehydes and ketones have **higher boiling points than alkanes** of similar relative formula masses. The reason for this is the presence of the polar carbonyl group which allows permanent dipole–permanent dipole attractions to be set up between their molecules. These are stronger and more difficult to break than the van der Waals' bonds which operate between alkane molecules.

Although aldehyde and ketone molecules can hydrogen bond with water molecules, no such bonds can exist between aldehyde molecules or ketone molecules.

Aldehydes and ketones, however, have **lower boiling points than alcohols** of similar relative formula masses because permanent dipole–permanent dipole attractions are weaker than the hydrogen bonds between the alcohol molecules. Aldehydes and ketones are generally insoluble in water, but the lower members can hydrogen bond with water molecules and dissolve to some extent.

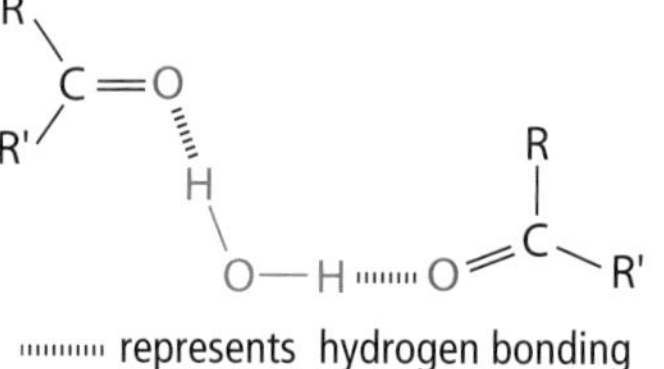

Lower aldehydes and ketones can dissolve in water by forming hydrogen bonds with water moledules.

REACTIONS OF ALDEHYDES AND KETONES

Aldehydes and ketones undergo:

- **reduction to primary and secondary alcohols** respectively by reaction with **lithium aluminium hydride** ($LiAlH_4$) dissolved in ether.

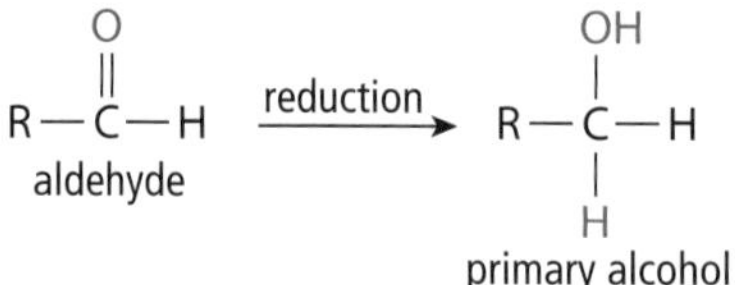

R—C(=O)—R' (ketone) —reduction→ R—CH(OH)—R' (secondary alcohol)

- **nucleophilic addition** with **hydrogen cyanide** to form **cyanohydrins**. In practice, hydrogen cyanide is not used directly since it is a highly poisonous gas. Instead, acidified potassium cyanide is used as the reagent. The reaction is described as **nucleophilic** addition since the **cyanide ion** (CN^-) is a **nucleophile**. It attacks the carbon atom of the carbonyl group since it has a slight positive charge due to oxygen having a greater electronegativity than carbon. The cyanohydrins that are produced can be hydrolysed to hydroxycarboxylic acids by refluxing them with dilute hydrochloric acid. For example, the cyanohydrin, 2-hydroxypropanenitrile, is hydrolysed to 2-hydroxypropanoic acid.

H_3C—CH=O (ethanal) + CN^- H^+ → H_3C—CH(CN)—OH (2-hydroxypropanenitrile (a cyanohydrin))

H_3C—CH(CN)—OH (2-hydroxypropanenitrile) —H_2O/H^+→ H_3C—CH(COOH)—OH (2-hydroxypropanoic acid)

- **a nucleophilic addition–elimination** reaction with **hydrazine** ($H_2N–NH_2$) to form **hydrazones**. Consider, for example, the reaction of propanone with hydrazine. To help explain the type of reaction taking place, we will now look at the mechanism.

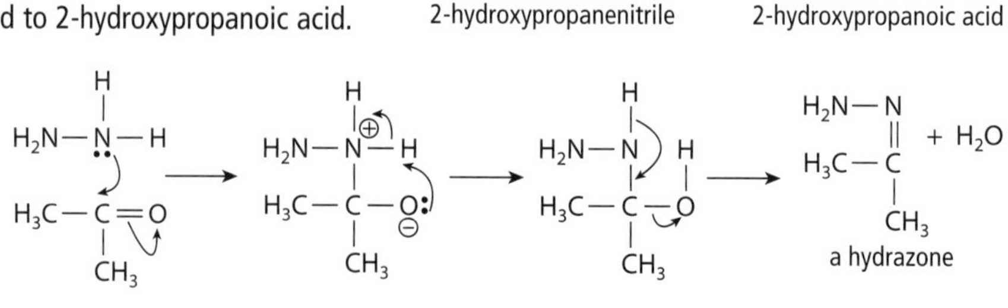

Initially, the lone pair of electrons on one of the N atoms in hydrazine attacks the slightly positively charged carbon atom of the carbonyl group in propanone. Following this nucleophilic attack, a proton (a hydrogen ion) is transferred from the nitrogen atom onto the oxygen atom. This is followed immediately by the elimination of a water molecule and the hydrazone is formed.

contd

REACTIONS OF ALDEHYDES AND KETONES contd

The overall equation for the reaction shows that **nucleophilic addition–elimination** can also be described as **condensation,** since water is formed when propanone and hydrazine react together.

$$H_3C-\underset{\displaystyle CH_3}{\underset{|}{C}}=O \quad \begin{matrix} H \\ H \end{matrix}\!\!>N-NH_2 \longrightarrow H_3C-\underset{\displaystyle CH_3}{\underset{|}{C}}=N-NH_2 + H_2O$$

A condensation reaction occurs when propanone and hydrazine join together, eliminating water in the process.

Aldehydes are generally **more reactive** when under nucleophilic attack than **ketones**. Ketones have two alkyl groups attached to the carbon atom of the polar carbonyl group. Because alkyl groups are electron-donating, the partial positive charge on this carbon atom is smaller in a ketone, making it less susceptible to attack by a nucleophile. In addition, the two alkyl groups in a ketone hinder the nucleophile's access to the carbon of the carbonyl group and this further reduces its reactivity.

DISTINGUISHING BETWEEN ALDEHYDES AND KETONES

Fehling's solution and **Tollens' reagent** can be used to distinguish an aldehyde from a ketone. Fehling's solution is alkaline and contains blue complexed copper(II) ions. Tollens' reagent is also alkaline but contains complexed silver(I) ions. When an aldehyde is heated with Fehling's solution, the blue colour disappears and a brick-red precipitate of copper(I) oxide is formed. In this process, the aldehyde is oxidised to a carboxylic acid while the copper(II) ions are reduced to copper(I) ions: $Cu^{2+}(aq) + e^- \rightarrow Cu^+(s)$

When an aldehyde is heated with Tollens' reagent, silver metal is precipitated, often as a silver mirror on the walls of the test tube. The aldehyde is again oxidised and the silver(I) ions are reduced to silver atoms: $Ag^+(aq) + e^- \rightarrow Ag(s)$

When ketones are heated with Fehling's solution or Tollens' reagent, no reaction takes place.

DON'T FORGET

When aldehydes or ketones react with hydrazine and with 2,4-dinitrophenylhydrazine, the products formed are called hydrazones and 2,4-dinitrophenylhydrazones respectively.

IDENTIFYING ALDEHYDES AND KETONES

When an aldehyde or ketone is added to **Brady's reagent** (a solution of 2,4-dinitrophenylhydrazine), an orange–yellow precipitate is formed, demonstrating the presence of the carbonyl group. A condensation or nucleophilic addition–elimination reaction takes place. The precipitate that is formed is called a 2,4-dinitrophenylhydrazone.

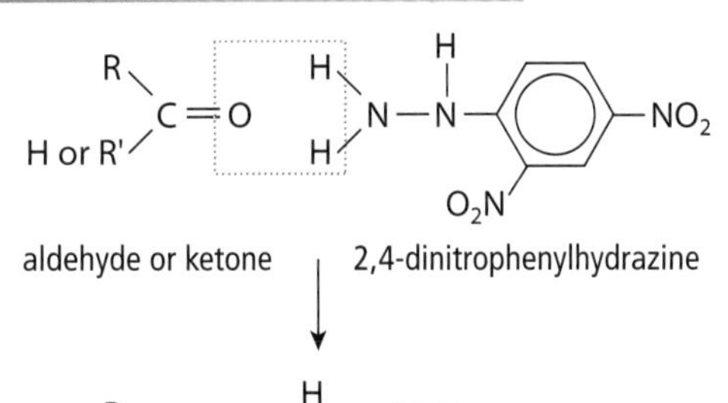

An unknown aldehyde or ketone can be identified by preparing its 2,4-dinitrophenylhydrazone derivative. The melting point of the derivative is determined and compared with the melting points of 2,4-dinitrophenylhydrazones of known compounds. The unknown aldehyde or ketone can thus be identified.

DON'T FORGET

Only aldehydes and ketones react with hydrogen cyanide, hydrazine and 2,4-dinitropenylhydrazine. Carboxylic acids and esters do not despite the presence of a carbonyl group within their structures.

LET'S THINK ABOUT THIS

Reducing sugars, like glucose, give a positive test with Fehling's solution or Tollens' reagent. This implies that they must contain an aldehyde group. In the solid state, glucose has the ring structure shown on the left and there is no sign of an aldehyde group. However, in aqueous solution the ring opens and an equilibrium is established between the ring structure and an open chain structure. You can see that the latter contains an aldehyde group (shown in red). This explains why glucose gives a positive test with both Fehling's solution and Tollens' reagent.

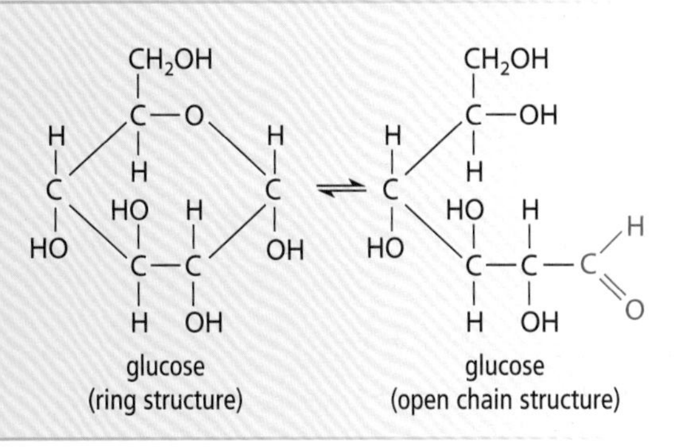

SYSTEMATIC ORGANIC CHEMISTRY 7 – CARBOXYLIC ACIDS AND AMINES

PHYSICAL PROPERTIES OF CARBOXYLIC ACIDS

$R{-}C(=O \cdots H{-}O)(O{-}H \cdots O=)C{-}R$

······ represents hydrogen bonding

In the pure state, carboxylic acids dimerise through the formation of hydrogen bonds.

DON'T FORGET

In pure carboxylic acids, hydrogen bonding allows molecules to pair up forming dimers, but dimerisation does not occur in aqueous solutions of these acids.

Carboxylic acids are either solids or liquids at room temperature. This reflects the extensive hydrogen bonding that exists between their molecules. As the carbon chain gets longer, hydrogen bonding diminishes. In the pure state, carboxylic acids tend to pair up and exist as **dimers**. They are able to do this through the formation of hydrogen bonds. No such dimers exist in aqueous solutions of carboxylic acids.

The lower members of the carboxylic acid series are soluble in water because they are able to form hydrogen bonds with water molecules. As the chain length of the carboxylic acid increases, water solubility decreases.

Carboxylic acids are **weak acids** because they are only partially ionised in aqueous solution. The carbon-to-oxygen bond lengths in the carboxylate ion are identical and lie between typical values for C–O single and C=O double bonds. This means that electrons are delocalised over the carbon and two oxygen atoms. This, in turn, provides the carboxylate ion with some stability and explains why carboxylic acids dissociate to a slight extent in water.

$$R{-}C(=O){-}O{-}H + H_2O \rightleftharpoons R{-}CO_2^{\ominus} + H_3O^{\oplus}$$

In carboxylic acids electrons are delocalised over the two C–O bonds.

PREPARATION OF CARBOXYLIC ACIDS

Carboxylic acids can be prepared by:

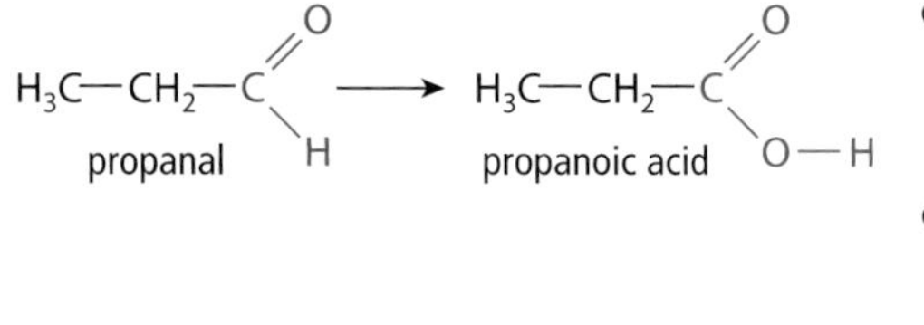

- **oxidising primary alcohols** or **aldehydes** by heating them under reflux with acidified potassium dichromate solution. For example, propanal can be oxidised to propanoic acid.
- **hydrolysing nitriles** (see page 72), **esters** or **amides** by heating them under reflux in the presence of a catalyst. Either an acid or an alkali can be used as catalyst in the two reactions shown here.

$$H_3C{-}C(=O){-}O{-}CH_3 + H_2O \rightarrow H_3C{-}C(=O){-}OH + HOCH_3$$

methyl ethanoate → ethanoic acid

$$H_3C{-}CH_2{-}C(=O){-}N(H){-}H + H_2O \rightarrow H_3C{-}CH_2{-}C(=O){-}OH + HNH_2$$

propanamide → propanoic acid

Esters and amides (and nitriles) can be used to make carboxylic acids.

REACTIONS OF CARBOXYLIC ACIDS

In aqueous solution, **carboxylic acids** behave as typical acids and form **salts** on reaction with:

- some **metals**, for example $Mg + 2CH_3COOH \rightarrow H_2 + Mg^{2+}(CH_3COO^-)_2$
- **carbonates**, for example $Na_2CO_3 + 2CH_3CH_2COOH \rightarrow CO_2 + H_2O + 2Na^+CH_3CH_2COO^-$
- **alkalis**, for example $KOH + HCOOH \rightarrow H_2O + K^+HCOO^-$

Carboxylic acids also undergo:

- **condensation** reactions with **alcohols** to form **esters**. The reactions are catalysed by concentrated phosphoric or sulphuric acids.

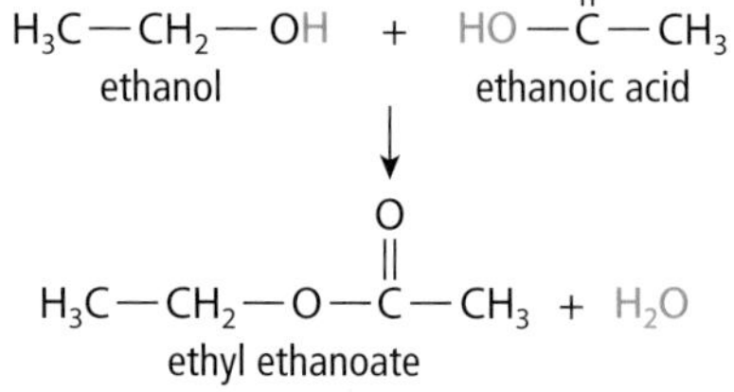

contd

REACTIONS OF CARBOXYLIC ACIDS contd

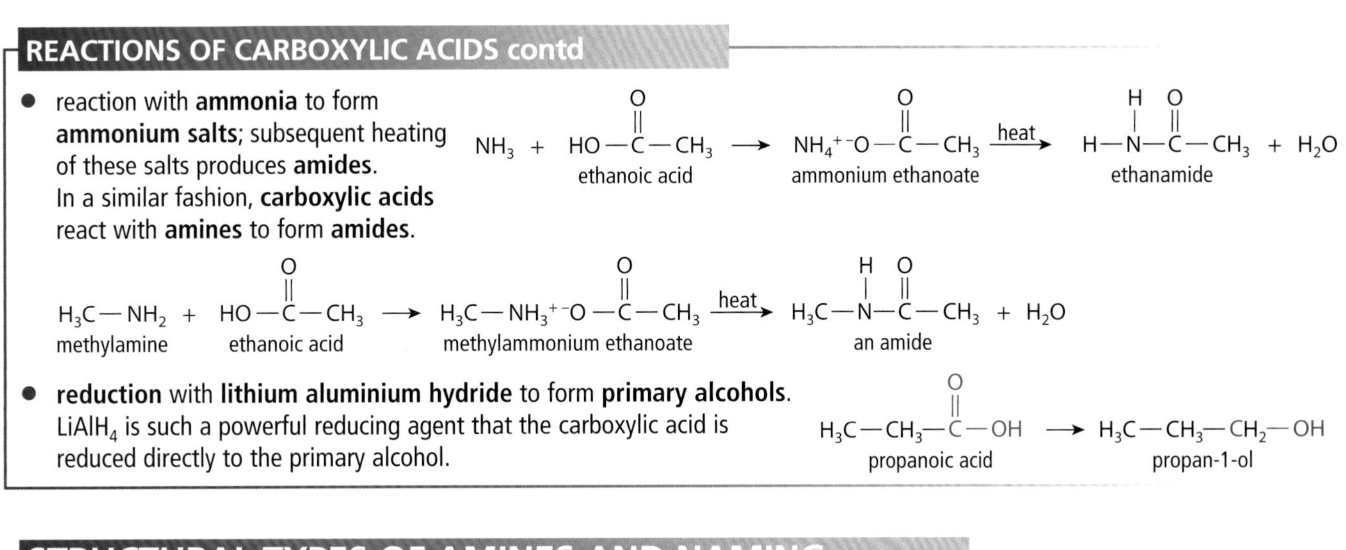

- reaction with **ammonia** to form **ammonium salts**; subsequent heating of these salts produces **amides**. In a similar fashion, **carboxylic acids** react with **amines** to form **amides**.

$$NH_3 + HO{-}C({=}O){-}CH_3 \rightarrow NH_4^{+}\,^{-}O{-}C({=}O){-}CH_3 \xrightarrow{heat} H{-}N(H){-}C({=}O){-}CH_3 + H_2O$$

ethanoic acid → ammonium ethanoate → ethanamide

$$H_3C{-}NH_2 + HO{-}C({=}O){-}CH_3 \rightarrow H_3C{-}NH_3^{+}\,^{-}O{-}C({=}O){-}CH_3 \xrightarrow{heat} H_3C{-}N(H){-}C({=}O){-}CH_3 + H_2O$$

methylamine + ethanoic acid → methylammonium ethanoate → an amide

- **reduction** with **lithium aluminium hydride** to form **primary alcohols**. $LiAlH_4$ is such a powerful reducing agent that the carboxylic acid is reduced directly to the primary alcohol.

$$H_3C{-}CH_2{-}C({=}O){-}OH \rightarrow H_3C{-}CH_2{-}CH_2{-}OH$$

propanoic acid → propan-1-ol

STRUCTURAL TYPES OF AMINES AND NAMING

Amines are **organic derivatives of ammonia** in which one or more hydrogen atoms in ammonia have been replaced by alkyl groups. There are three structural types of amines – primary, secondary and tertiary. They are classified according to the number of alkyl groups attached to the nitrogen atom.

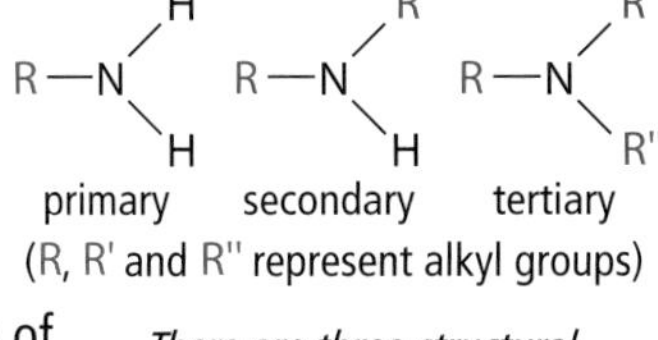

The most common method of **naming amines** is to prefix the word 'amine' with the names of the alkyl groups (arranged in alphabetical order) attached to the nitrogen atom. For example, $\mathbf{CH_3CH_2NHCH_3}$ is called **ethylmethylamine** and $\mathbf{(CH_3)_3N}$ is called **trimethylamine**.

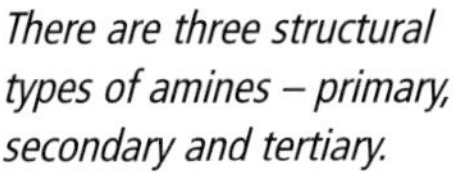

There are three structural types of amines – primary, secondary and tertiary.

PHYSICAL PROPERTIES OF AMINES

Primary and secondary amines contain a polar N–H bond and so have **hydrogen bonds** between their molecules. No such bonds can be set up between molecules of tertiary amines, since they do not contain a hydrogen atom directly bonded to the highly electronegative nitrogen atom. This explains why primary and secondary amines have **higher boiling points** than their isomeric tertiary amines.

Amines with low relative formula masses are soluble in water because they can form hydrogen bonds with water molecules. This is true even for tertiary amines, as can be seen in the diagram.

||||||||| represents hydrogen bonding

Even tertiary amines can form hydrogen bonds with water molecules, although they cannot form hydrogen bonds between their own molecules, unlike primary and secondary amines.

Amines, like ammonia, are **weak bases** and dissociate to a slight extent in aqueous solution. In the reaction, the lone pair of electrons on the nitrogen atom in the amine molecule accepts a proton (hydrogen ion) from the water molecule, thus generating alkylammonium ions and hydroxide ions. The latter ions make the solution alkaline. For example,

$$CH_3NH_2(aq) + H_2O(l) \rightleftharpoons CH_3NH_3^{+}(aq) + OH^{-}$$

methylammonium ion

Reactions of amines

Amines react with

- **mineral acids**, like hydrochloric, sulphuric and nitric acids, to form **salts**.

$$CH_3CH_2NH_2 + HNO_3 \rightarrow CH_3CH_2NH_3^{+}NO_3^{-}$$

ethylammonium nitrate

- **carboxylic acids** to form **salts**. On heating these salts, water is lost and amides are formed as shown above.

DON'T FORGET

Primary, secondary and tertiary amines can hydrogen bond with water molecules but only primary and secondary amines have hydrogen bonds between their molecules.

LET'S THINK ABOUT THIS

Which of the following isomeric amines has the lowest boiling point?

A $C_4H_9NH_2$ **B** $C_3H_7NHCH_3$ **C** $C_2H_5NHC_2H_5$ **D** $C_2H_5N(CH_3)_2$

SYSTEMATIC ORGANIC CHEMISTRY 8 – AROMATICS

STRUCTURE AND BONDING IN BENZENE

Benzene (C_6H_6) is the simplest member of the class of hydrocarbons called aromatic hydrocarbons or arenes. In the planar benzene molecule, each carbon atom is **sp^2 hybridised** and the three half-filled sp^2 hybrid orbitals form **sigma (σ) bonds** with a hydrogen atom and two neighbouring carbon atoms. This leaves an electron occupying a p orbital on each carbon atom. Each of these p orbitals overlaps side-on with the two p orbitals on either side and a **pi (π) molecular orbital** forms. The benzene molecule therefore contains **12 σ bonds** (black) and **one π molecular orbital** (red).

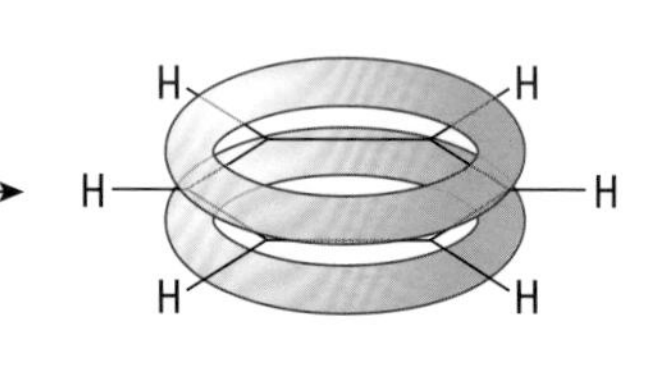

The p orbitals above and below each carbon atom (as shown on the left-hand diagram) overlap to form a pi orbital that extends above and below the carbon ring (as shown in the right-hand diagram).

The six electrons that occupy the pi molecular orbital are not tied to any one carbon atom but are shared by all six. They are **delocalised** and represented by a ring in the skeletal formula of benzene.

The unusual bonding and delocalisation of electrons found in benzene is represented by the structure shown here.

DON'T FORGET

Bonding in benzene can be described in terms of sp^2 hybridisation, sigma bonds and a pi molecular orbital containing delocalised electrons.

REACTIONS OF BENZENE

Benzene is unusually stable and it is the delocalised electrons which account for this stability. The delocalised electrons also explain why benzene tends to undergo substitution reactions rather than addition reactions. Addition reactions would disrupt the delocalisation and so reduce the stability of the ring. Substitution reactions, on the other hand, can occur without any such disruption and the stability of the benzene ring is maintained. It is the delocalised electrons in the pi molecular orbital which make benzene susceptible to attack by electrophiles (electron-pair acceptors). As a result, benzene undergoes **electrophilic substitution reactions**.

Some typical electrophilic substitution reactions of benzene

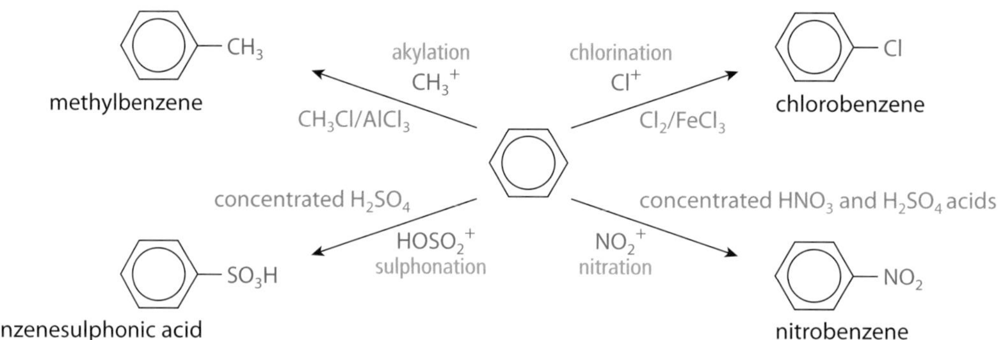

Reactions of benzene: the electrophiles are shown in red, the reagents in blue and the reaction names in green.

As well as **chlorination**, benzene can undergo **bromination** when bromine is used as the reagent along with aluminium chloride or iron(III) chloride as catalyst. The electrophiles, Cl^+ and Br^+, are generated by the reaction between the halogen and the catalyst:

$Cl_2 + AlCl_3$ or $FeCl_3 \rightarrow Cl^+ + [AlCl_4]^-$ or $[FeCl_4]^-$

$Br_2 + AlCl_3$ or $FeCl_3 \rightarrow Br^+ + [AlCl_3Br]^-$ or $[FeCl_3Br]^-$

The electrophile in the **nitration** of benzene is the **nitronium ion**, NO_2^+, which is generated by the reaction between the concentrated nitric and sulphuric acids:

$HNO_3 + 2H_2SO_4 \rightarrow NO_2^+ + 2HSO_4^- + H_3O^+$

contd

REACTIONS OF BENZENE contd

In the **sulphonation** of benzene, the electrophile, $HOSO_2^+$, is generated from the concentrated sulphuric acid. $H_2SO_4 + H_2SO_4 \rightarrow HOSO_2^+ + HSO_4^- + H_2O$

Oleum, which is concentrated sulphuric acid with sulphur trioxide added, can also be used as the reagent and the electrophile in this case is SO_3.

When benzene is **alkylated**, it reacts with a **halogenoalkane** in the presence of aluminium chloride as catalyst. In the example above, chloromethane is used and the electrophile CH_3^+ is generated from its reaction with the aluminium chloride.
$CH_3Cl + AlCl_3 \rightarrow CH_3^+ + [AlCl_4]^-$

DON'T FORGET

Benzene resists addition reactions but undergoes electrophilic substitution reactions.

PHENOL AND PHENYLAMINE

Phenol is an aromatic compound with the abbreviated structural formula shown here. C_6H_5–OH

It contains an **–OH group** just like alcohols, but unlike alcohols, phenol is a **weak acid** and dissociates to a slight extent in aqueous solution.

$C_6H_5OH + H_2O \rightleftharpoons C_6H_5O^- + H_3O^+$

phenoxide ion

Phenol is a weak acid in water.

The reason why phenol dissociates is due to the stability of the **phenoxide ion**. This stability arises from the fact that an electron in a p orbital on the oxygen atom becomes part of the pi molecular orbital of the benzene ring. Electron delocalisation, therefore, extends over the whole ion (and not just the benzene ring) and so increases its stability. Phenol is a weaker acid than the carboxylic acids but is more strongly acidic than the alcohols. Being an acid, phenol will form salts on reaction with alkalis.

$C_6H_5OH + NaOH \rightarrow C_6H_5O^-Na^+ + H_2O$

sodium phenoxide

Phenol neutralises sodium hydroxide solution forming the salt, sodium phenoxide.

Phenylamine, also known as **aniline** and **aminobenzene**, is an **aromatic primary amine**.

C_6H_5–NH_2

Phenylamine is an aromatic primary amine.

Like the aliphatic amines discussed on page 75, phenylamine is a **weak base** and partially ionises in aqueous solution. In the process, the lone pair of electrons on the nitrogen atom accepts a proton (a hydrogen ion) from the water molecule and the phenylammonium ion is formed along with a hydroxide ion.

$C_6H_5NH_2 + H_2O \rightleftharpoons C_6H_5NH_3^+ + OH^-$

phenylammonium ion

Phenlyamine is a weak base that partially ionises in solution to give a phenylammonium ion and a hydroxide ion.

In general, the degree of weakness of an amine depends on how ready the lone pair of electrons on the nitrogen atom is to accept a proton. The less ready it is, the weaker the amine. In the case of phenylamine, the lone pair of electrons on the nitrogen atom becomes part of the pi molecular orbital of the benzene ring; electron delocalisation is extended into the amino group. This makes the phenylamine molecule more stable than the phenylammonium ion where no such extension of electron delocalisation can take place. As a result, the lone pair of electrons on the nitrogen atom is unwilling to accept a proton. So, phenylamine is a very weak base and considerably weaker than any aliphatic amines. Despite its weakly basic nature, phenylamine reacts with mineral acids to form salts.

$C_6H_5NH_2 + HCl \rightarrow C_6H_5NH_3^+Cl^-$

phenylammonium chloride

When phenylamine is added to hydrochloric acid, the salt phenylammonium chloride is formed.

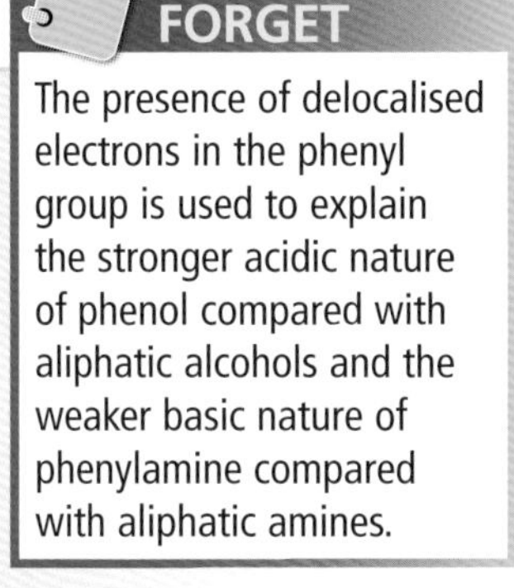

DON'T FORGET

The presence of delocalised electrons in the phenyl group is used to explain the stronger acidic nature of phenol compared with aliphatic alcohols and the weaker basic nature of phenylamine compared with aliphatic amines.

LET'S THINK ABOUT THIS

N-phenylethanamide can be prepared from benzene in three steps.

benzene $\xrightarrow{\text{step 1}}$ $C_6H_5NO_2$ (nitrobenzene) $\xrightarrow{\text{step 2}}$ $C_6H_5NH_2$ (phenylamine) $\xrightarrow{\text{step 3}}$ $C_6H_5NHCOCH_3$ (N-phenylethanamide)

(a) What chemicals are required to react with benzene to bring about step 1?

(b) Name a reagent that could be used to bring about step 3.

(c) Name the type of reaction taking place in **(i)** step 1, **(ii)** step 2, **(iii)** step 3?

(d) Explain why phenylamine is a weaker base than methylamine.

ORGANIC CHEMISTRY

STEREOISOMERISM

ISOMERISM

Molecules which have the same molecular formula but differ in the way their atoms are arranged are called **isomers**. They are distinct compounds often with different physical and chemical properties.

There are two ways in which atoms can be arranged differently in isomers:

- the atoms are bonded together in a different order in each isomer – these are called **structural isomers**
- the order of bonding in the atoms is the same but the arrangement of the atoms in space is different for each isomer – these are called **stereoisomers**.

You have already covered **structural isomerism** in your earlier work in Chemistry and we will now explore **stereoisomerism**. There are two types of stereoisomerism – **geometric isomerism** and **optical isomerism**.

GEOMETRIC ISOMERISM

To illustrate what is meant by geometric isomerism, let's consider the alkene, but-2-ene: $H_3C–HC=CH–CH_3$. It contains a carbon-to-carbon double bond and the diagram shows the spatial arrangement of bonds round each carbon atom. This arrangement of atoms and bonds is planar and all the bond angles are 120°. Furthermore, the bonds are fixed in relation to one another. This means that it is impossible to rotate one end of an alkene molecule around the C=C double bond while the other end is fixed. This is one of the reasons why some alkenes can exhibit geometric isomerism.

cis-but-2-ene

trans-but-2-ene

Geometric isomers of but-2-ene; 'cis' means 'on the same side' and you can see that in cis-but-2-ene both methyl groups lie on the same side of the double bond. 'Trans' means 'on different sides' and so in trans-but-2-ene the methyl groups are on different sides of the double bond.

Returning to our example, you will find that there are two geometric isomers of but-2-ene. One is referred to as the ***cis* isomer** and the other, the ***trans* isomer**, as shown on the left. It is important to remember that geometric isomers are different compounds and have distinct physical properties. For example, *cis*-but-2-ene melts at –139°C while *trans*-but-2-ene melts at –105°C. But-2-enedioic acid also has *cis* and *trans* forms.

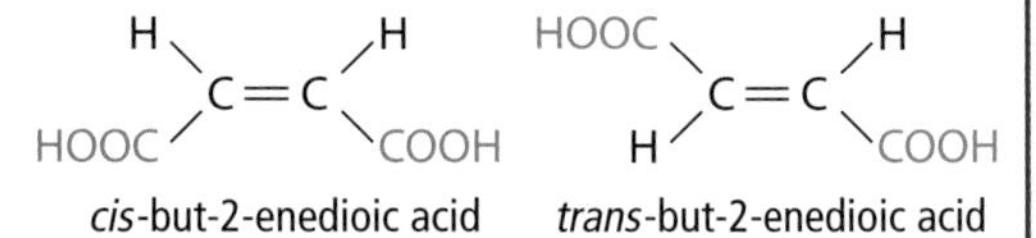

cis-but-2-enedioic acid *trans*-but-2-enedioic acid

As well as differing in their physical properties, *cis*-but-2-enedioic acid and *trans*-but-2-enedioic acid also differ in one of their chemical properties. *Cis*-but-2-enedioic acid is readily dehydrated to form a cyclic anhydride. However, since the carboxyl groups are on opposite sides of the C=C double bond in the *trans* isomer, they are not in a suitable orientation to undergo such a reaction.

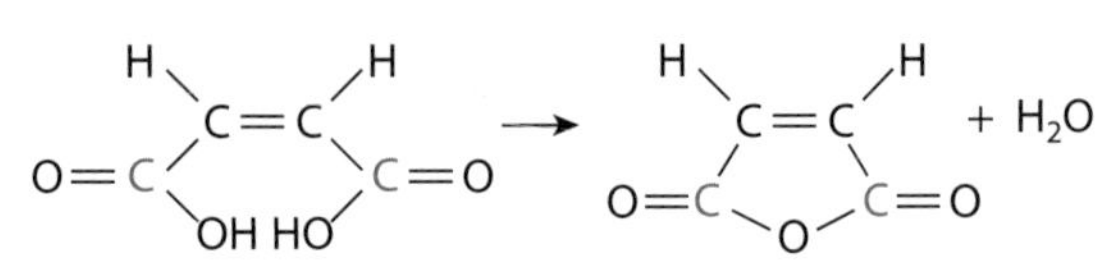

In the cis isomer the carboxyl groups are in the correct orientation to allow this dehydration to proceed; in the trans isomer they are not.

So, geometric isomerism can occur in organic compounds which contain a C=C double bond. However, in addition to the double bond, the molecule must have two different groups attached to each of the carbon atoms of the double bond. Propene, for example, would not exhibit geometric isomerism because it has two identical hydrogen atoms attached to one of the carbon atoms of the double bond.

Consider why propene does not form geometric isomers.

Geometric isomerism can arise due to the lack of free rotation around a bond, frequently a carbon-to-carbon double bond.

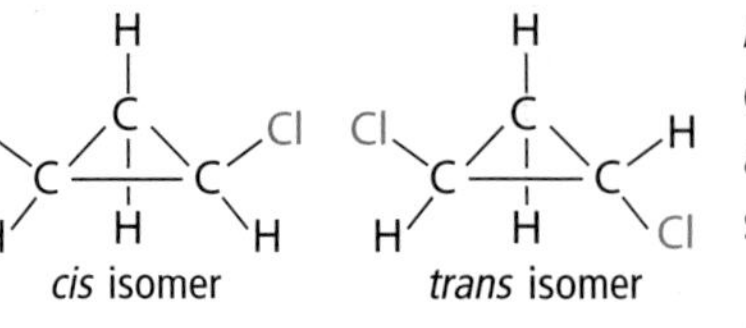
cis isomer *trans* isomer

Although geometric isomerism is most common in compounds containing a C=C double bond, it can also arise in saturated rings where rotation about the C–C single bonds is restricted.

1,2-dichlorocyclopropane has these two geometric isomers.

OPTICAL ISOMERISM

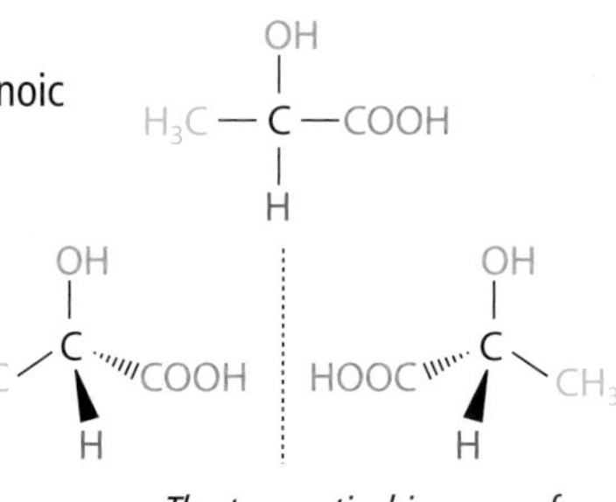

The two optical isomers of lactic acid.

To illustrate what is meant by optical isomerism, let's consider lactic acid (2-hydroxypropanoic acid). It contains a tetrahedral carbon atom (shown in black) with **four different groups** attached. As a result, lactic acid exists as two distinct isomeric forms.

Each isomer is the **mirror image** of the other; they are known as **optical isomers** or **enantiomers**. But all molecules have mirror images, yet they don't all exhibit optical isomerism. What makes lactic acid different is that its two isomers are **non-superimposable**. You should make molecular models of these optical isomers to convince yourself that one isomer can't be superimposed on the other.

In general, if a molecule contains a tetrahedral carbon atom that has four different groups attached to it, then it will have an optical isomer. These molecules can be described as **chiral**. You are already familiar with chiral objects in everyday life, for example hands, feet, golf clubs and so on.

Plane-polarised light

Optical isomers are identical in every physical property except their effect on plane-polarised light. A beam of light consists of an infinite number of waves vibrating in all planes perpendicular to the direction in which the light travels. If this beam of light is passed through a polariser (such as the polaroid film in polaroid sunglasses), all the vibrations are cut out except those in one plane. The light emerging from the polariser is plane-polarised light.

When plane-polarised light is passed through a solution containing one optical isomer, the plane of the polarised light is rotated through a certain angle. If this solution is replaced by an equimolar (same concentration) solution of the other optical isomer, it too rotates the plane of polarised light by exactly the same angle but in the opposite direction. For example, if one isomer of an optically active compound rotates the plane of polarised light by +50° (that is in a clockwise direction), the other optical isomer will rotate it by –50° (in an anticlockwise direction), provided the concentrations of the two solutions are equal. An equimolar mixture of the two optical isomers would have no effect on plane-polarised light since the rotational effect of one would be cancelled out by the opposite rotational effect of the other. Such a mixture is **optically inactive** and is known as a **racemic mixture**.

Optical isomerism can occur in compounds in which four different groups are arranged tetrahedrally around a carbon atom. The optical isomers have identical physical properties except for their effect on plane-polarised light.

Many of the organic compounds found in nature are chiral. More importantly, most natural compounds in living organisms are not only chiral, but are present in only one of their optical isomeric forms. Such compounds include amino acids, proteins, enzymes and sugars.

LET'S THINK ABOUT THIS

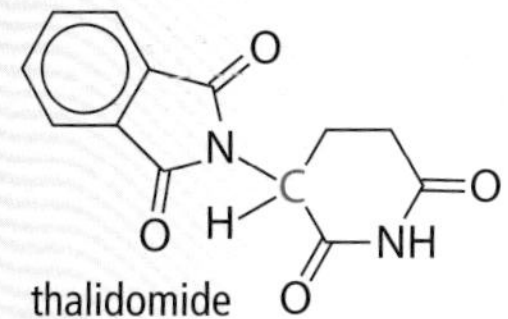

You may have heard of the '**thalidomide**' tragedy. Thalidomide, a drug, was widely prescribed in the 1950s and early 1960s to pregnant women in order to combat 'morning sickness'. Tragically, many of the babies born to these women had malformed limbs and it was later proved that thalidomide was linked to these birth defects.

Look at the structure of thalidomide shown above; you will spot a tetrahedral carbon atom attached to four different groups – it is the one coloured red. This implies that thalidomide exhibits optical isomerism and its two enantiomers are shown below.

The drug was administered as a racemic mixture; both optical isomers were present in equal proportions. The isomer labelled R prevented morning sickness while that labelled S caused the birth defects. Despite the bad reputation of thalidomide, it has since proved useful in alleviating a variety of disorders such as rheumatoid arthritis, leprosy, AIDS, tuberculosis and transplant rejection.

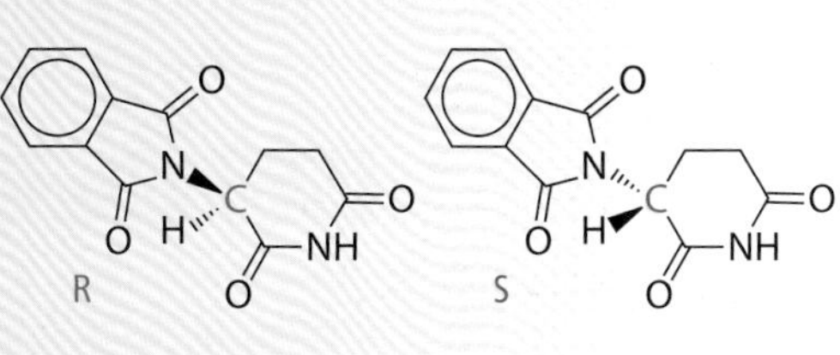

http://ochem.jsd.claremont.edu/tutorials.htm go to website and click on 'Optical Activity'

STRUCTURAL ANALYSIS 1

ELEMENTAL MICROANALYSIS

Elemental microanalysis (or **combustion analysis)** can be used in determining the **empirical formula** of an organic compound. The empirical formula shows the **simplest** whole number ratio of the different atoms in a compound. Take ethane (C_2H_6), for example. Its C:H ratio is 2:6, i.e. 1:3, and so its empirical formula is $\mathbf{CH_3}$.

C, H, N, S in sample

$\downarrow O_2$

$CO_2(g) + H_2O(g) + N_2(g) + SO_2(g)$

In modern combustion analysers, a tiny sample (approximately 2 mg) is accurately weighed and oxidised at high temperature in an atmosphere of oxygen. The product mixture of CO_2, H_2O, N_2 and SO_2 is separated by gas chromatography and the mass of each component is measured using a thermal conductivity detector. From these product masses, the mass of each of the elements C, H, N and S in the sample can be derived. If oxygen is present in the sample, its mass can be determined by subtracting the total mass of the other elements present from the mass of the original sample. The empirical formula can be calculated from the element masses using the method outlined below.

DON'T FORGET

Elemental microanalysis is used to determine the masses of the elements present in a sample of an organic compound in order to work out its empirical formula.

Calculating the empirical formula

Suppose complete combustion of 1·75 mg of an organic compound produced 3·51 mg of CO_2 and 1·43 mg of H_2O. No other products were formed.

1 mol CO_2 (44·0 g) contains 1 mol C (12·0 g)

$$\text{mass of C in sample} = 3{\cdot}51 \times 10^{-3} \times \frac{12{\cdot}0}{44{\cdot}0} = 9{\cdot}57 \times 10^{-4}\,\text{g}$$

1 mol H_2O (18·0 g) contains 2 mol H (2·0 g)

$$\text{mass of H in sample} = 1{\cdot}43 \times 10^{-3} \times \frac{2{\cdot}0}{18{\cdot}0} = 1{\cdot}59 \times 10^{-4}\,\text{g}$$

$$\text{mass of O in sample} = 1{\cdot}75 \times 10^{-3} - 9{\cdot}57 \times 10^{-4} - 1{\cdot}59 \times 10^{-4} = 6{\cdot}34 \times 10^{-4}\,\text{g}$$

Element	C	H	O
Mass (g)	$9{\cdot}57 \times 10^{-4}$	$1{\cdot}59 \times 10^{-4}$	$6{\cdot}34 \times 10^{-4}$
Number of moles	$\frac{9{\cdot}57 \times 10^{-4}}{12{\cdot}0} = 7{\cdot}98 \times 10^{-5}$	$\frac{1{\cdot}59 \times 10^{-4}}{1{\cdot}0} = 1{\cdot}59 \times 10^{-4}$	$\frac{6{\cdot}34 \times 10^{-4}}{16{\cdot}0} = 3{\cdot}96 \times 10^{-5}$
Mole ratio	$\frac{7{\cdot}98 \times 10^{-5}}{3{\cdot}96 \times 10^{-5}} = 2{\cdot}02$ (2)	$\frac{1{\cdot}59 \times 10^{-4}}{3{\cdot}96 \times 10^{-5}} = 4{\cdot}02$ (4)	$\frac{3{\cdot}96 \times 10^{-5}}{3{\cdot}96 \times 10^{-5}} = 1{\cdot}00$ (1)

Empirical formula = C_2H_4O

The molecular formula of the compound can be found if you know its relative molecular mass. Suppose it was 88.

$$\frac{\text{Relative molecular mass}}{\text{Relative empirical mass}} = \frac{88}{2(12{\cdot}0) + 4(1{\cdot}0) + 1(16{\cdot}0)} = \frac{88}{44} = 2$$

So, the molecular formula is $C_4H_8O_2$, twice the empirical formula.

MASS SPECTROMETRY

Mass spectrometry is a technique used in determining the **accurate molecular mass** and **structural features of an organic compound**.

In mass spectrometry, a minute sample (approximately 1×10^{-9} g) of the unknown organic compound is vapourised and injected into the mass spectrometer where it is bombarded by high-energy electrons. The energy is sufficient to knock electrons out of the molecules and, as a result, they ionise and break into smaller ion fragments.

contd

MASS SPECTROMETRY contd

These **positively charged ions**, mostly with a 1+ charge, are accelerated by a high-voltage electric field into a strong magnetic field which deflects them into a series of separate ion paths according to their mass/charge (m/z) ratio. Positive ions with lower mass/charge ratios are deflected more than those with higher ratios. Each separated ion path is detected and a spectrum is recorded.

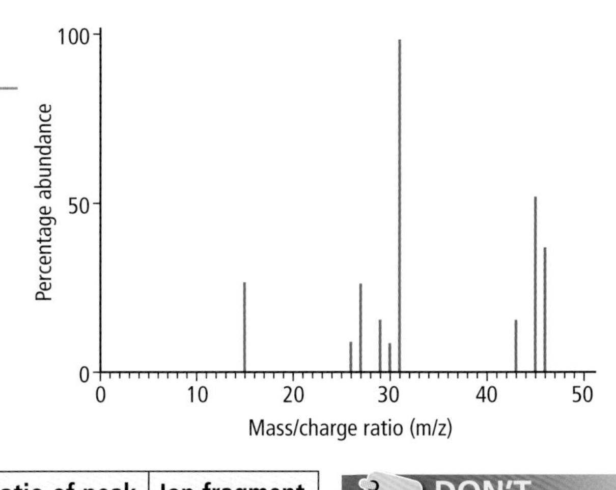

The **mass spectrum** shown is that for **ethanol** (CH_3CH_2OH). The peak with the highest m/z ratio provides the molecular mass of the organic compound. In the example, this appears at m/z = 46 and confirms the molecular mass of ethanol as 46. This peak arises from the so-called **molecular ion**, $[CH_3CH_2OH]^+$ (the molecule with one electron removed). The table shows the formula of the molecular ion and the formulae of the ion fragments that are responsible for some of the other peaks in the spectrum.

m/z ratio of peak	Ion fragment
15	$[CH_3]^+$
29	$[CH_3CH_2]^+$
31	$[CH_2OH]^+$
45	$[CH_3CH_2O]^+$
46	$[CH_3CH_2OH]^+$

DON'T FORGET

You must always include the positive charge when writing the structure of an ion fragment or molecular ion.

X-RAY CRYSTALLOGRAPHY

X-ray crystallography is a technique used to determine the precise three-dimensional structure of an organic compound. As the name suggests, a single **crystal** of the organic compound is exposed to **X-rays** of a single wavelength. Since the interatomic distances in a compound are approximately equal to the wavelength of the X-rays, the crystal acts as a diffraction grating. The X-rays are scattered by the electrons in the atoms making up the organic molecules and, from the diffraction pattern, electron-density contour maps of the molecule, like the one shown in the diagram, can be constructed. From these maps, the identity and precise spatial arrangement of each atom in the molecule, apart from hydrogen atoms, can be determined.

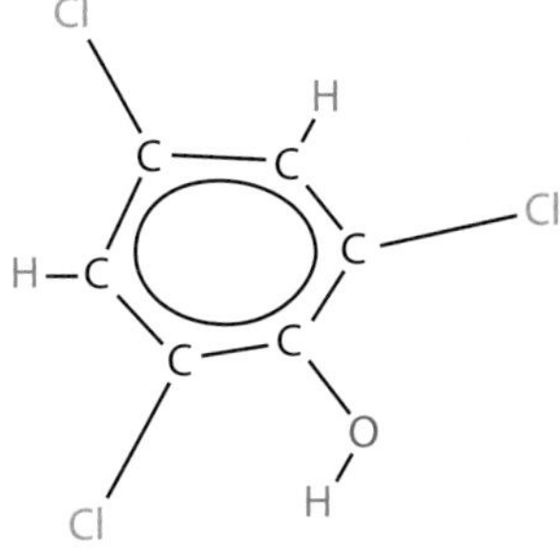

This electron-density contour map is obtained from TCP (2,4,6-trichlorophenol) and it has been overlayed with the structure of TCP. You will notice that the Cl, C and O atoms are much more clearly resolved than the H atoms. This results from the fact that these atoms have far more electrons to diffract the X-rays than an H atom.

In the early days, a structure determination by X-ray crystallography was limited to small molecules and could take years. Modern methods have reduced this time to a matter of hours and very large molecules, like proteins, are now routinely analysed.

LET'S THINK ABOUT THIS

Although mass spectrometry is used principally to elucidate the structures of organic compounds, it has also been used to establish reaction mechanisms. Consider, for example, the hydrolysis of the ester, ethyl ethanoate, to form ethanol and ethanoic acid.

When it reacts with water, which of the C–O bonds is broken? Is it the one marked in red or the one marked in blue?

When water containing ^{18}O ($H_2{}^{18}O$) is used, the mass spectrum of the products shows that

- the molecular ion for the ethanol product occurs at m/z = 46 indicating that no ^{18}O has been taken up into the ethanol molecule
- the molecular ion for the ethanoic acid product occurs at m/z = 62 indicating that ^{18}O has been taken up into the ethanoic acid molecule.

This establishes that the hydrolysis of an ester involves breaking the red C–O bond, that is the one adjacent to the carbonyl group.

http://www.chemguide.co.uk/analysis/masspec/howitworks.html Mass spectrometry

STRUCTURAL ANALYSIS 2

INFRA-RED SPECTROSCOPY

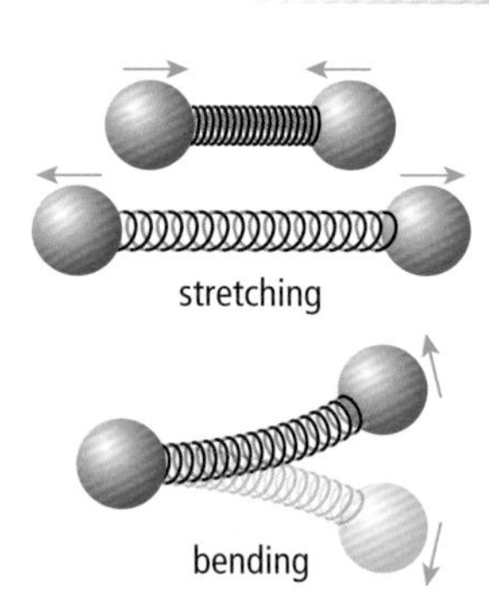

A useful model of a bond, whereby it is represented as a spring joining two atoms, is shown in the diagram. There are a number of different types of bond vibration and two of these are illustrated here. They are referred to as stretching and bending.

DON'T FORGET

The absorption of infra-red radiation causes bonds within a molecule to vibrate and infra-red spectroscopy can be used to identify functional groups in an organic molecule.

Infra-red radiation comprises that part of the electromagnetic spectrum that lies between microwaves and visible light (see page 4). When it is absorbed by organic compounds, the energy is sufficient to cause **bonds** within the molecules to **vibrate,** but not enough to break the bonds.

The wavelength of the infra-red radiation that is absorbed when a bond vibrates depends on the type of atoms which make up the bond and the stiffness (strength) of the bond. In general, light atoms joined by stiff bonds absorb radiation of shorter wavelength (higher energy) than heavier atoms joined by looser bonds. **Infra-red spectroscopy** makes use of these characteristics and is an important analytical tool; it can be used to identify certain **bonds and functional groups** in organic molecules.

An infra-red spectrum can be obtained for a sample of an organic compound regardless of its physical state – solid, liquid, gas or dissolved in a solvent. In the spectrometer, infra-red radiation is passed through the sample. Some wavelengths are absorbed, causing bond vibrations within the molecules. The transmitted radiation then passes to a detector where the intensity at different wavelengths is measured. An infra-red spectrum, like that shown below, results.

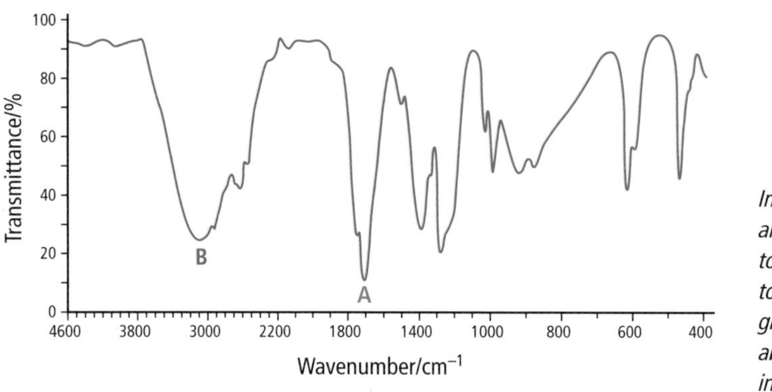

In an infra-red spectrum, an absorption corresponds to a 'peak' pointing towards the bottom of the graph. Notice too that the absorptions are expressed in terms of wavenumber.

Absorptions are given in terms of **wavenumber**. Wavenumber is the reciprocal of wavelength,

wavenumber $= \frac{1}{\text{wavelength}} = \frac{1}{\lambda}$ and the **units** are normally **cm^{-1}**.

This particular infra-red spectrum is that for ethanoic acid.

ethanoic acid: $H_3C{-}C(=O){-}O{-}H$

Absorption peak A, at 1700 cm^{-1}, is due to stretching of the C=O bond. The broad peak B, at 2600–3400 cm^{-1}, corresponds to stretching of the hydrogen bonded O–H bond in the acid.

Since particular types of vibration always occur at a similar wavenumber, it is possible to build up a table of characteristic absorptions. Such a table is to be found on page 13 of the SQA Data Booklet. If you examine this table, you will see, for example, that an absorption in the wavenumber range 2260–2215 cm^{-1} is indicative of a nitrile and is due to stretching of the C≡N bond. So, given the infra-red spectrum of an unknown organic compound and a table of characteristic absorptions, it should be possible to identify the functional groups present in the compound. In most cases, however, more information is required to determine the full structure.

NUCLEAR MAGNETIC RESONANCE SPECTROSCOPY

Some atomic nuclei can spin about their own axes. For example, an ^1H nucleus (a **hydrogen nucleus** or **proton**) can spin in one of two directions – clockwise and anticlockwise. As a result, protons behave as tiny magnets and, when placed between the poles of a powerful magnet, some align themselves with the field of the magnet while others align against it.

contd

NUCLEAR MAGNETIC RESONANCE SPECTROSCOPY contd

Those protons aligned with the field have a slightly lower energy than those aligned against it. The energy difference between the two states corresponds to the radio-frequency region of the electromagnetic spectrum. So, when protons are exposed to radio waves, energy is absorbed to promote those in the lower energy state to the higher energy state. In effect, the protons 'flip' from being aligned with the magnetic field to being aligned against it. As the protons fall back to the lower energy state, the same radio frequency that was absorbed is emitted. This can be measured with a radio receiver. This phenomenon is known as **proton nuclear magnetic resonance (proton NMR)**.

Proton NMR spectroscopy is a powerful analytical tool and gives information about:

- the different chemical environments of the protons in an organic molecule
- how many protons are in each of these environments.

Take ethanol, for example. You can see from its structural formula that the protons are in three different chemical environments: the H nuclei in the CH_3 group, the H nuclei in the CH_2 group and the H nucleus in the OH group.

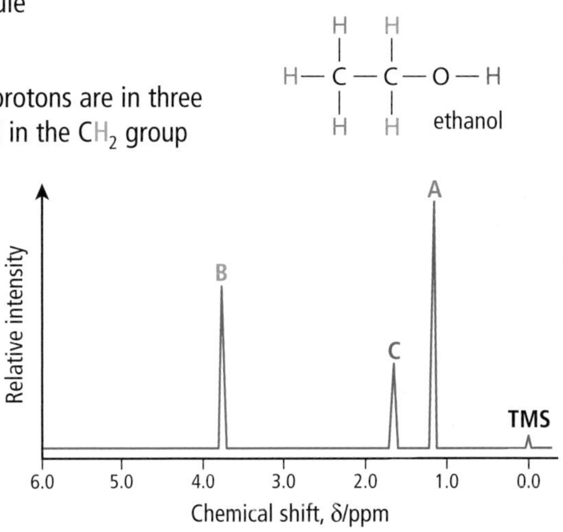

The low-resolution NMR spectrum for ethanol is shown.

The three peaks A, B and C correspond to the three different chemical environments of the protons. The area under each peak is proportional to the number of protons in that environment. In this case, the areas under peaks A, B and C are in the ratio **3:2:1** corresponding to the **three** protons in the CH_3 group, the **two** protons in the CH_2 group and the **one** proton in the OH group.

Notice the small peak labelled TMS. **TMS** is short for tetramethylsilane, $Si(CH_3)_4$, and it is used as a standard against which all absorptions due to other proton environments are measured. TMS is assigned a value of zero and the difference between the protons in TMS and the protons in other chemical environments is known as the **chemical shift,** which is given the symbol, δ. The chemical shift is measured in parts per million (ppm). Chemical shift values for protons in different chemical environments are given on page 15 of the SQA Data Booklet.

If you examine this chart, you will notice that protons in

- a CH_3 group have a chemical shift in the range 0·9–1·5 ppm and so peak A is due to the CH_3 protons in ethanol
- a CH_2 group in an alcohol have a chemical shift in the range 3·5–3·9 ppm and so peak B is due to the CH_2 protons in ethanol
- an OH group in an alcohol have a chemical shift in the range 1·0–5·0 ppm and so peak C must be due to the OH proton in ethanol.

A proton nuclear magnetic resonance spectrum provides information about the different chemical environments of hydrogen nuclei (protons) in an organic molecule and the relative numbers of protons in each of these environments.

LET'S THINK ABOUT THIS

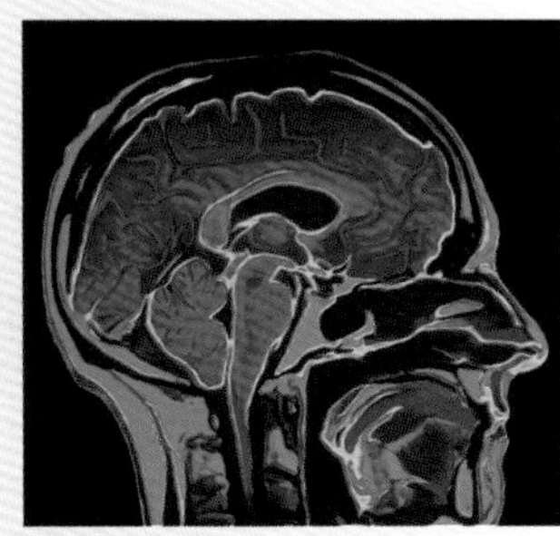

Since hydrogen is the most common element in the human body, it is not surprising that nuclear magnetic resonance found application in the field of medicine as **magnetic resonance imaging** or **MRI**.

NMR imaging in medicine is based on the time it takes for hydrogen nuclei (protons) in the unstable high-energy state to 'relax' or return to the low-energy state. These relaxation times are different for protons in fat, muscle (proteins), blood and bone because of differences in their chemical environments. The relaxation times are enhanced by computer, to produce the magnetic resonance image (as seen here). A major advantage of MRI scanning over X-rays is that the patient is exposed to radio-frequency radiation, avoiding the damage caused by X-rays.

MEDICINES

INTRODUCTION

Medicines are those drugs that have a beneficial effect on the body.

There is a misconception that drugs and medicines are quite different. The term 'drug' carries with it the connotation of addiction, abuse and crime but, in fact, medicines are just a subset of drugs. **Drugs** are defined as substances that can alter the biochemical processes in the body and **medicines** are those drugs which have a beneficial effect.

HISTORICAL DEVELOPMENT

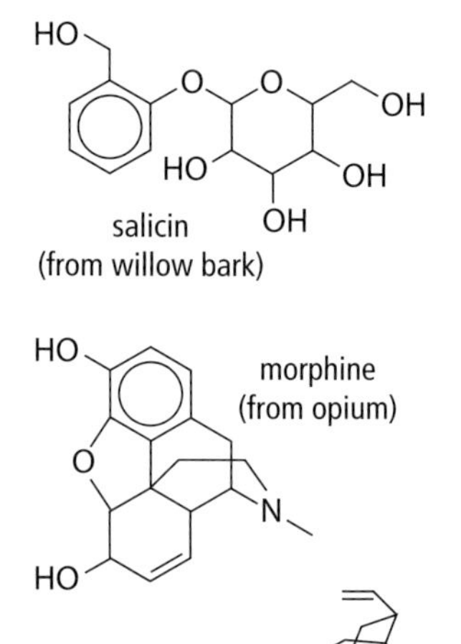

The skeletal structures of the pharmacologically active compounds found in plant materials are shown in the diagrams above.

Since ancient times, the curative powers of brews or potions derived from certain plants have been recognised. Examples include:

- the bark of the willow tree used to relieve pain, lower fever and to reduce inflammation
- opium used as a sedative and pain-killer
- the bark of the cinchona tree used to treat malaria.

It wasn't until the 1800s that the pharmacologically active compounds in some of these extracts were isolated and identified.

As chemical knowledge and understanding improved, chemists were able to determine the structures of some of these pharmacologically active compounds. This lead on to the synthesis of the active compounds themselves, and to the synthesis of their derivatives. As a result, the range of effective medicines rapidy expanded.

Take aspirin, for example. It has its origins in salicin, the active ingredient in willow bark. When salicin is hydrolysed it produces glucose and salicyl alcohol; the latter can be oxidised to salicylic acid. Salicylic acid was also used as a medicine but it caused irritation and bleeding in the stomach and intestines. However, when salicylic acid is treated with ethanoic anhydride, acetylsalicylic acid (**aspirin**) is produced.

Nowadays, computers play an important role in the design and development of medicines. For example, molecular modelling software allows chemists to study how potential medicines will interact with receptors in the body and to determine how their structures can be modified to enhance their effect.

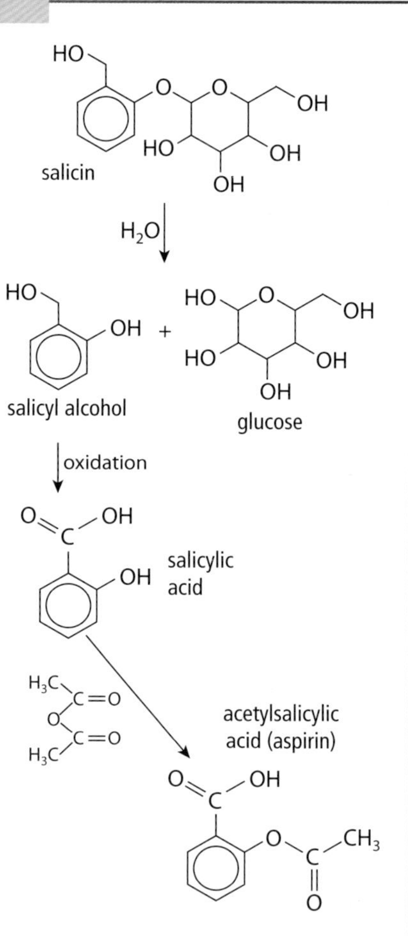

Aspirin has proved to be a very effective medicine with few side effects, unlike salicylic acid which caused stomach irritation.

HOW DO MEDICINES WORK?

Most medicines work by binding to **receptors** within the body, which are normally **protein molecules**. Some of these proteins are embedded in the membranes that surround cells, while others are located inside the cells. The latter are **enzymes** (globular proteins) and, since they act as catalysts in the body, they are referred to as **catalytic receptors**.

Membrane-embedded protein receptors have hollows or clefts on their surfaces into which small biologically active molecules can fit and bind. Once in the **binding site**, the active molecule triggers a response. The binding sites in catalytic receptors are similar but are usually referred to as **active sites** and the molecules which bind to these sites are normally called **substrate** molecules. The catalytic receptors catalyse a reaction on these substrate molecules.

contd

HOW DO MEDICINES WORK? contd

The shape of the molecule that binds to the receptor site is critical – it must complement the shape of the receptor site. When the molecule fits into the receptor site, the binding forces hold it in place so that it does not float away. These binding forces can be hydrogen bonds, van der Waal's forces and even ionic bonds.

Agonist and antagonist medicines

Most medicines can be classified as **agonists** or **antagonists** according to the response they trigger when bound to a receptor site. An **agonist** mimics the body's naturally active molecule and, on binding to the receptor site, it triggers the same response as the natural molecule. An **antagonist**, on the other hand, binds more strongly to the receptor site and prevents the naturally active molecule from binding. As a result, the antagonist molecule blocks the site and so the natural response is not triggered.

An agonist medicine triggers a response like the body's naturally active compound, while an antagonist medicine produces no response by preventing the action of the body's naturally active compound.

An example of an **agonist medicine** is **salbutamol**, which is used in the treatment of asthma. Asthma attacks are caused when the bronchioles (airways in the lungs) narrow and become blocked with mucus. In such circumstances, the body responds by releasing adrenaline, which binds to receptors and triggers the dilation of the bronchioles. Unfortunately, it also triggers an acceleration of the heart rate and an increase in blood pressure, which could cause a heart attack. The medicine salbutamol binds more strongly to the receptor sites than adrenaline and, while it triggers the widening of the bronchioles, it does not trigger an increase in heart rate. So, salbutamol relieves asthma attacks without the risk of a heart attack.

An example of an **antagonist medicine** is **propranolol,** which is used to treat high blood pressure. As we know, the natural compound adrenaline, when bound to receptor sites, triggers an increase in blood pressure. The medicine propranolol binds preferentially to the same sites, but does not trigger any response. So, the propranolol blocks these sites preventing the action of adrenaline.

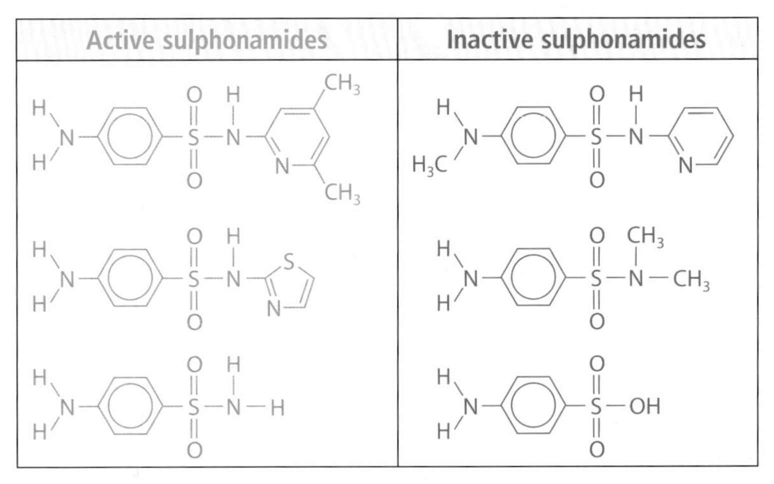

The structural fragment of a medicine that confers its pharmacological activity is called the **pharmacaphore**. Let's consider the structural formulae of some sulphonamides and see if we can identify the pharmacaphore. Some sulphonamides are antibacterial agents (shown in green in the first column of the table) and some (in the second column in red) have no antibacterial activity.

By examining both sets of structures in the table, you can see that the fragment that makes a sulphonamide active (the pharmacaphore) has the structure shown here.

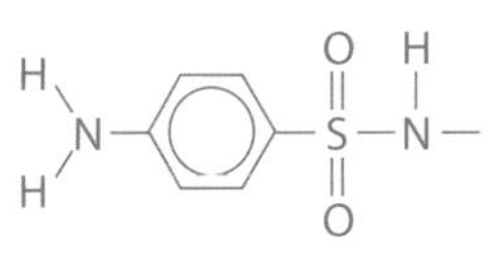

LET'S THINK ABOUT THIS

Sulphanilamide is an enzyme inhibitor that blocks the biosynthesis of folic acid, which is essential for cell growth. It does this by mimicking 4-aminobenzoic acid, one of the reactants required in folic acid synthesis.

sulphanilamide

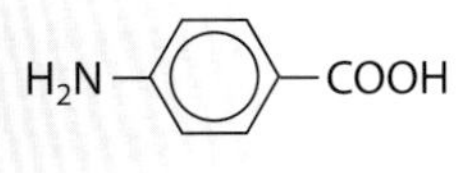

4-aminobenzoic acid

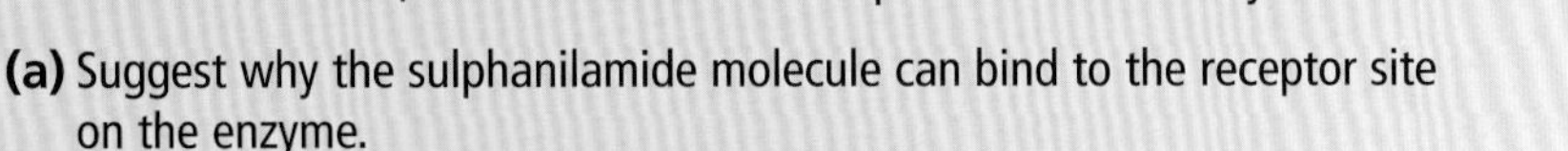

(a) Suggest why the sulphanilamide molecule can bind to the receptor site on the enzyme.

(b) State whether sulphanilamide plays the role of agonist or antagonist and explain your answer.

http://www.authorstream.com/Presentation/neilcredo-87474-receptors-agonists-antagonists-education-ppt-powerpoint/

UNIT 3 PPAs 1–5

DON'T FORGET

This is a dehydration reaction and not a condensation reaction because the water molecule has been removed from just one reactant molecule and not two.

UNIT 3 PPA 1 – PREPARATION OF CYCLOHEXENE

Aim

To prepare cyclohexene from cyclohexanol and determine the percentage yield.

Introduction

Cyclohexene can be prepared by dehydrating cyclohexanol using concentrated phosphoric acid. The cyclohexene can be separated from the reaction mixture by distillation and, after purification, it can be weighed and the percentage yield calculated.

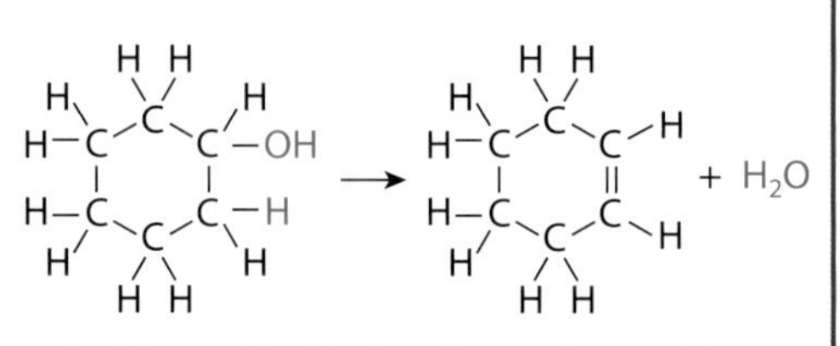

Cyclohexanol is dehydrated to produce cyclohexene.

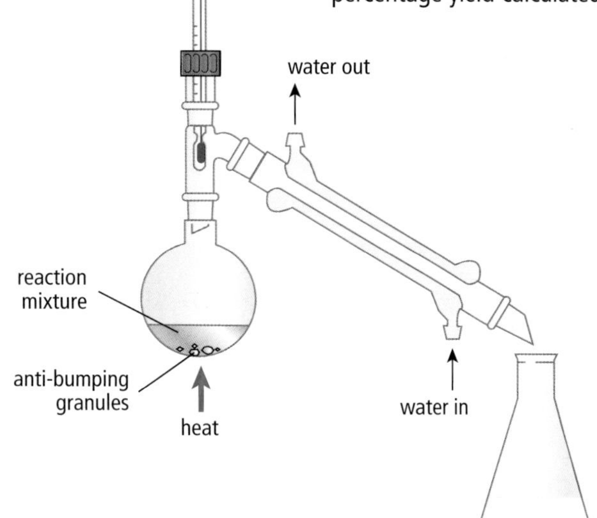

Procedure

About 20 g of cyclohexanol was added to a pre-weighed round-bottomed flask, and the flask and contents were reweighed accurately. Approximately 8 cm^3 of concentrated phosphoric acid was added dropwise to the cyclohexanol. After adding a few anti-bumping granules, the reaction mixture was slowly distilled using the apparatus shown.

The liquid that came over between 70°C and 90°C was collected. The distillate was poured into a separating funnel and about an equal volume of saturated sodium chloride solution was added. The separating funnel was shaken thoroughly and left to stand, allowing the two layers to separate. The lower aqueous layer was run off and disposed of. The upper layer – the crude cyclohexene – was run off into a small conical flask containing a few pieces of anhydrous calcium chloride, a drying agent. The dried cyclohexene was then decanted into a round-bottomed flask together with a few anti-bumping granules. This crude cyclohexene was slowly distilled and the liquid which came over between 80°C and 85°C was collected in a pre-weighed receiver flask. The flask which now contained pure cyclohexene was reweighed accurately.

A few drops of the pure cyclohexene were added to bromine solution to test for unsaturation.

DON'T FORGET

When a reaction mixture is heated, there is a tendency for it to boil violently as large bubbles of superheated vapour suddenly erupt from the mixture. This is prevented by the addition of the anti-bumping granules.

Results

Mass of round-bottomed flask	= 30·46 g
Mass of round-bottomed flask + cyclohexanol	= 50·25 g
Mass of cyclohexanol	= 19·79 g
Mass of receiver flask	= 33·96 g
Mass of receiver flask + cyclohexene	= 42·48 g
Mass of cyclohexene	= 8·52 g

DON'T FORGET

In order to calculate the percentage yield, you need to know the accurate mass of the limiting reactant (the one which is not in excess), and the accurate mass of the pure product that is formed.

From the balanced equation:

1 mol cyclohexanol → 1 mol cyclohexene

100·0 g ⟷ 82·0 g

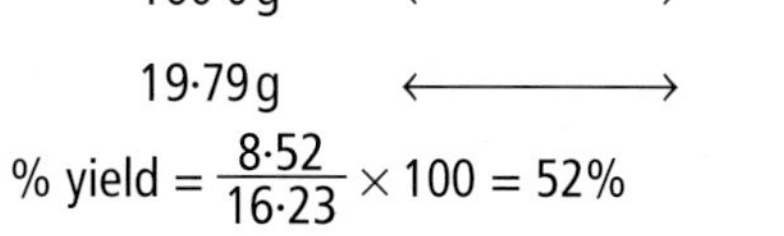

$$19{\cdot}79\text{ g} \longleftrightarrow 19{\cdot}79 \times \frac{82{\cdot}0}{100{\cdot}0} = 16{\cdot}23\text{ g}$$

$$\%\text{ yield} = \frac{8{\cdot}52}{16{\cdot}23} \times 100 = 52\%$$

The cyclohexene that was prepared in this experiment was a colourless liquid with a boiling point of 80–85°C. It turned the orange-red bromine solution colourless.

contd

UNIT 3 PPA 1 – PREPARATION OF CYCLOHEXENE contd

Conclusions

A sample of cyclohexene was prepared and the percentage yield was 52%.

The bromine solution was decolourised, indicating that the cyclohexene is unsaturated.

Evaluation

The yield of cyclohexene was less than 100% because:

- cyclohexene would be lost during its transfer from one container to another
- cyclohexene is volatile and appreciable amounts would be lost through evaporation
- there may have been impurities present in the cyclohexanol.

Concentrated phosphoric acid was used to dehydrate the cyclohexanol rather than concentrated sulphuric acid because it gives a higher yield of cyclohexene. With concentrated sulphuric acid more side reactions occur and it tends to produce extensive charring. The purpose of shaking the crude cyclohexene with saturated sodium chloride solution rather than water was to maximise the yield of cyclohexene. Saturated sodium chloride solution is more polar than water and less cyclohexene dissolves in it. In addition, saturated sodium chloride solution is more dense than water and so separates from the cyclohexene more rapidly.

To cut down loss of the volatile cyclohexene during distillation, the receiver flask could have been placed in an ice bath.

UNIT 3 PPA 2 – IDENTIFICATION BY DERIVATIVE FORMATION

Aim

To identify an unknown liquid ketone by making a derivative of it using 2,4-dinitrophenylhydrazine and then comparing the melting point of the derivative with those of known derivatives.

Introduction

It is difficult to identify any unknown liquid from its boiling point because boiling points vary significantly with atmospheric pressure. Melting points of solids, on the other hand, do not vary with atmospheric pressure and are fixed. So, by converting the liquid ketone into a solid derivative the problem with atmospheric pressure can be overcome. In this experiment, the unknown ketone is treated with 2,4-dinitrophenylhydrazine solution (Brady's reagent) and a solid derivative known as a 2,4-dinitrophenylhydrazone is formed by a condensation reaction.

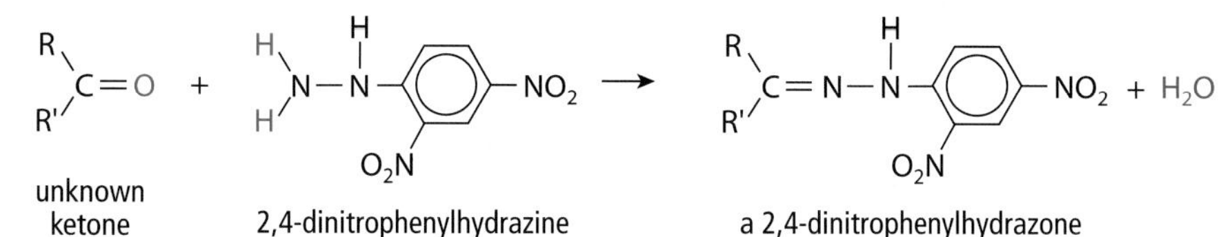

The solid derivative is purified by recrystallisation. In such a procedure, the crude derivative is dissolved by heating in a minimum volume of a suitable solvent. The hot saturated solution is filtered to remove any insoluble impurities and then allowed to cool. Crystals of the pure derivative appear and are all filtered off. Soluble impurities are left behind in the filtrate or mother-liquor as it is often called. The choice of solvent is critical. It must not react with the derivative and, ideally, the derivative should be highly soluble in the hot solvent and virtually insoluble in the cold solvent.

The melting point of the pure derivative is then measured and compared with the melting points of known ketone derivatives (see the table on page 88) in order to identify the unknown ketone.

DON'T FORGET

When an aldehyde or ketone reacts with 2,4-dinitrophenylhydrazine, the product formed is called a 2,4-dinitrophenylhydrazone.

DON'T FORGET

When recrystallising a substance, the solvent chosen must not react with the substance. The substance should be much more soluble in the hot solvent than it is in the cold solvent.

contd

UNIT 3 PPA 2 – IDENTIFICATION BY DERIVATIVE FORMATION contd

Ketone	Melting point of 2,4-dinitrophenyhydrazone/°C	Ketone	Melting point of 2,4-dinitrophenyhydrazone/°C
propanone	128	pentan-3-one	156
butanone	115	hexan-2-one	107
pentan-2-one	141	cyclohexanone	162

DON'T FORGET

The melting point of the unknown ketone derivative is compared with the melting points of known ketone derivatives and **not** the ketones themselves.

Procedure

About 10 cm^3 of 2,4-dinitrophenylhydrazine solution (Brady's reagent) was added to a test tube followed by approximately 10 drops of the unknown ketone. The test tube was stoppered and inverted several times to ensure thorough mixing. The orange precipitate was filtered off under pressure and transferred to a conical flask containing about 10 cm^3 of ethanol and a few anti-bumping granules. The mixture was gently heated until the solid dissolved. Then it was filtered through a pre-heated filter paper and funnel, into a conical flask. The solution was allowed to cool to room temperature and crystals of the pure derivative appeared. These were filtered off, washed with a little cold ethanol and dried in an oven at about 50–60°C. Finally, a melting point apparatus was used to determine the melting point of the crystals.

Results

Melting point of derivative = 125–127°C

The crystals of the derivative were long, needle-shaped and orange in colour.

Conclusion

By comparing the melting point of the derivative with the melting points of known derivatives shown in the table above, it can be concluded that the unknown ketone was propanone.

Evaluation

DON'T FORGET

When measuring the melting point of a substance, it is vitally important to raise the temperature of the block inside the melting point apparatus very slowly, otherwise the melting point will be underestimated.

The melting point of the derivative was sharp, melting over a narrow range of 2°C, and was very close to the accepted value of 128°C. Both of these characteristics indicate that the derivative was pure.

The warm solution of the derivative was filtered through a hot rather than a cold filter paper and funnel in order to reduce the risk of crystals separating out on the filter paper and in the stem of the funnel.

When using the melting point apparatus it was important to raise the temperature very, very slowly. With a rapid rise in temperature, the melting point of the derivative would have been underestimated. This is because the mercury in the thermometer takes time to respond to rising temperature of the heating block. As a result, the thermometer reading lags behind the temperature of the block and the more rapid the heating rate, the wider the gap between the two.

UNIT 3 PPA 3 – PREPARATION OF BENZOIC ACID BY HYDROLYSIS OF ETHYL BENZOATE

Aims

To prepare benzoic acid from ethyl benzoate, calculate the percentage yield and determine the melting point of the benzoic acid sample.

Introduction

The ethyl benzoate is first hydrolysed to benzoic acid and ethanol; this reaction is catalysed by the sodium hydroxide. The sodium hydroxide then neutralises the benzoic acid to form sodium benzoate.

Benzoic acid can be prepared by the alkaline hydrolysis of the ester, ethyl benzoate.

$$C_6H_5-C(=O)-O-CH_2CH_3 + NaOH \rightarrow C_6H_5-C(=O)-O^-Na^+ + HOCH_2CH_3$$

ethyl benzoate → sodium benzoate

contd

UNIT 3 PPA 3 – PREPARATION OF BENZOIC ACID BY HYDROLYSIS OF ETHYL BENZOATE contd

By adding the strong acid, hydrochloric acid, to the reaction mixture, the weak benzoic acid is displaced.

$$C_6H_5COO^-Na^+(aq) + HCl(aq) \longrightarrow C_6H_5COOH(s) + Na^+Cl^-(aq)$$

sodium benzoate → benzoic acid

Benzoic acid is displaced by the addition of hydrochloric acid.

The crude benzoic acid is separated by filtration and recrystallised from water. The pure sample of benzoic acid is weighed and the percentage yield is calculated. The melting point of the pure benzoic acid can also be determined.

Procedure

About 5 g of ethyl benzoate was added to a pre-weighed round-bottomed flask and the flask and contents were reweighed accurately. Approximately 50 cm^3 of 2 mol l^{-1} sodium hydroxide solution and a few anti-bumping granules were added to the ethyl benzoate. The reaction mixture was heated under reflux using the apparatus shown. Once the oily drops had disappeared, heating was stopped.

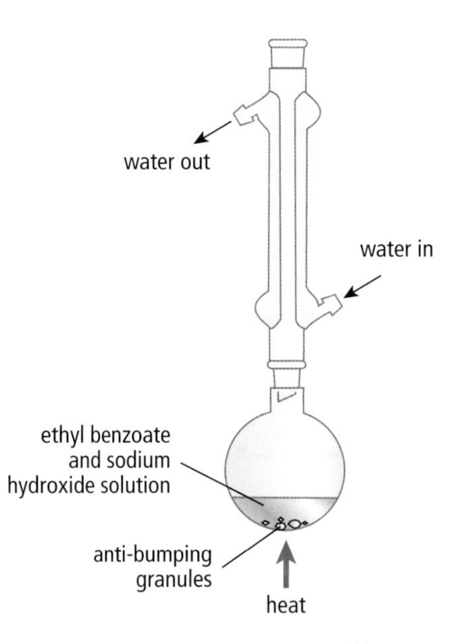

After cooling, the reaction mixture was transferred to a beaker. 5 mol l^{-1} hydrochloric acid was slowly added, with stirring, until the mixture turned blue litmus paper red. The white precipitate of benzoic acid that had formed was filtered off and recrystallised from water. The pure benzoic acid crystals were then filtered off, transferred to a pre-weighed clock glass and dried in an oven at about 70°C. The clock glass and dry benzoic acid crystals were then reweighed accurately.

The melting point of the benzoic acid was then determined.

Results

Mass of round-bottomed flask	= 40·25 g
Mass of round-bottomed flask + ethyl benzoate	= 45·61 g
Mass of cyclohexanol	= 5.36 g
Mass of clock glass	= 9·62 g
Mass of clock glass + benzoic acid	= 12·86 g
Mass of benzoic acid	= 3·24 g

From the balanced equations:

1 mol ethyl benzoate → 1 mol benzoic acid

150·0 g ⟷ 122·0 g

5·36 g ⟷ $5{\cdot}36 \times \frac{122{\cdot}0}{150{\cdot}0} = 4{\cdot}36$ g

$\%\ \text{yield} = \frac{3{\cdot}24}{4{\cdot}36} \times 100 = 74\%$

Melting point of benzoic acid = 119–120°C

The benzoic acid crystals were long, white and needle-shaped.

Ethyl benzoate, like most esters, is insoluble in water and so the disappearance of the oily drops indicates that the ester has undergone hydrolysis.

Conclusions

A sample of benzoic acid was prepared and the percentage yield was 74%. The melting point of benzoic acid was sharp, melting over a narrow range of 1°C, and was close to the accepted value of 122°C. This evidence indicated that the sample was pure.

Evaluation

The yield of benzoic acid was less than 100% because:

- benzoic acid crystals are lost during their transfer from one container to another
- benzoic acid is slightly soluble in water and in the recrystallisation process some of it would remain dissolved in the water.

Water is a suitable solvent in the recrystallisation process because benzoic acid does not react with water and is highly soluble in hot water and only slightly soluble in cold water.

UNIT 3 PPA 3 – PREPARATION OF BENZOIC ACID BY HYDROLYSIS OF ETHYL BENZOATE contd

Had the hydrolysis of ethyl benzoate been carried out using hydrochloric acid rather than sodium hydroxide, the yield of benzoic acid would have been considerably reduced. The hydrolysis of ethyl benzoate to benzoic acid and ethanol is a reversible reaction and reaches a state of equilibrium. Both hydrochloric acid and sodium hydroxide catalyse the reaction but sodium hydroxide, unlike hydrochloric acid, goes on to react with the benzoic acid produced. This drives the equilibrium to the products' side and so increases the yield of benzoic acid.

UNIT 3 PPA 4 – PREPARATION OF ASPIRIN

Aims

To prepare aspirin from 2-hydroxybenzoic acid, calculate the percentage yield and determine the melting point of the aspirin sample.

Introduction

Aspirin (acetylsalicylic acid) can be prepared by the condensation or esterification reaction between 2-hydroxybenzoic acid (salicylic acid) and ethanoic anhydride.

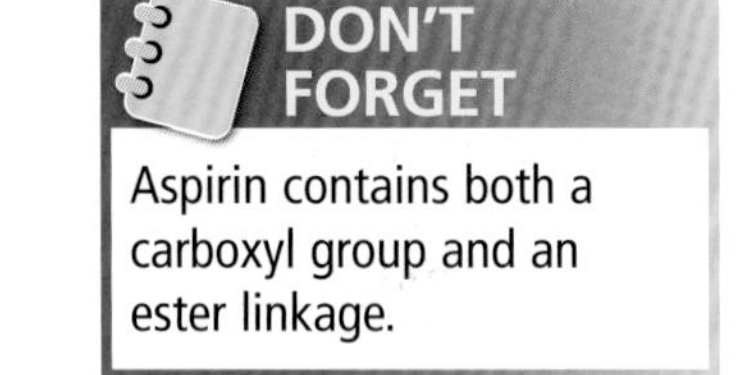

Aspirin contains both a carboxyl group and an ester linkage.

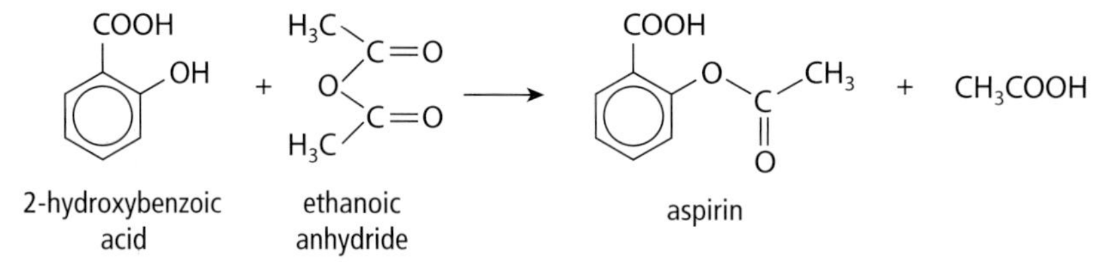

The crude aspirin is separated by filtration and recrystallised. The pure sample of aspirin is weighed and the percentage yield is calculated. The melting point of the pure aspirin can also be determined.

The concentrated phosphoric acid was added to the reaction mixture to catalyse the reaction.

Procedure

About 5 g of 2-hydroxybenzoic acid was added to a pre-weighed conical flask and the flask and contents were reweighed accurately. 10 cm^3 of ethanoic anhydride (an excess) was added from a measuring cylinder to the 2-hydroxybenzoic acid along with a few drops of concentrated phosphoric acid.

The reaction mixture was heated on a hot plate to approximately 85°C. This temperature was held for about 10 minutes and the mixture was constantly stirred. The flask and contents were cooled in an ice-bath and the reaction mixture was then poured into about 150 cm^3 of cold water. The white precipitate of crude aspirin that formed was filtered off, washed with several portions of cold water and then transferred into about 15 cm^3 of ethanol contained in a conical flask. The mixture was heated gently until the crude aspirin dissolved and then poured into a beaker containing about 40 cm^3 of cold water. This mixture was set aside to cool and white crystals of pure aspirin formed. The pure aspirin crystals were filtered off, transferred to a pre-weighed clock glass and dried in an oven at about 100°C. The clock glass and dry aspirin crystals were then reweighed accurately.

The melting point of the aspirin was then determined.

Results

Mass of conical flask = 38·39 g
Mass of conical flask + 2-hydroxybenzoic acid = 43·45 g
Mass of 2-hydroxybenzoic acid = 5·06 g
Mass of clock glass = 10·86 g
Mass of clock glass + benzoic acid = 14·06 g
Mass of benzoic acid = 3·20 g

contd

UNIT 3 PPA 4 – PREPARATION OF ASPIRIN contd

From the balanced equation:

1 mol 2-hydroxybenzoic acid $\rightarrow$ 1 mol aspirin

138·0 g $\longleftrightarrow$ 180·0 g

5·06 g $\longleftrightarrow$ $5{\cdot}06 \times \frac{180{\cdot}0}{138{\cdot}0} = 6{\cdot}60\,\text{g}$

$\%\ \text{yield} = \frac{3{\cdot}20}{6{\cdot}60} \times 100 = 48\%$

Melting point of aspirin = 129–133°C

The aspirin crystals were white and consisted of short, thin rods.

The ethanoic anhydride was known to be in excess and so the percentage yield calculation is based on the mass of the limiting reactant, 2-hydroxybenzoic acid.

Conclusions

A sample of aspirin was prepared and the percentage yield was 48%. The melting point of benzoic acid was not sharp, melting over a fairly wide range of 4°C, and it was quite a few degrees below the accepted value of 137°C. This evidence indicated that the sample was not completely pure.

Evaluation

The yield of aspirin was less than 100% because:

- aspirin crystals would be lost during their transfer from one container to another
- aspirin is slightly soluble in ethanol and, in the recrystallisation process, some of it would remain dissolved in the ethanol–water mix.

The –OH group in 2-hydroxybenzoic acid is attached to a benzene ring and this makes it weakly acidic (see page 77).This means that it is less reactive to esterification with ethanoic acid than the –OH group in alcohols. So, to esterify the –OH group in 2-hydroxybenzoic acid, a reagent more reactive than ethanoic acid is needed and this is why ethanoic anhydride is used.

The low melting point of the crystals indicated that they were not completely pure. A possible impurity that could be present is unreacted 2-hydroxybenzoic acid.

Another reagent that could be used in place of ethanoic anhydride is ethanoyl chloride, CH_3COCl (see page 70).

UNIT 3 PPA 5 – ASPIRIN DETERMINATION

Aims

To determine the mass of aspirin in a tablet.

Introduction

Since it is insoluble in water, aspirin cannot be determined by direct titration. An indirect method such as back titration must be used.This involves treating a sample of aspirin of accurately known mass with a known but excess amount of sodium hydroxide.The alkali first catalyses the hydrolysis of aspirin to ethanoic acid and salicylic acid (2-hydroxybenzoic acid) and then neutralises these acids. The overall balanced equation for the reaction is:

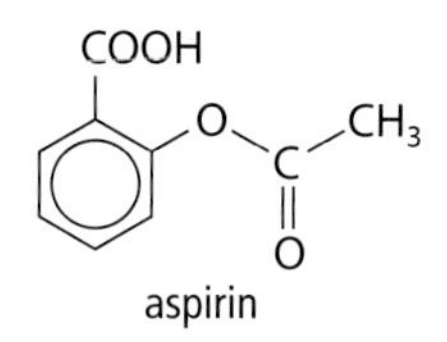

aspirin

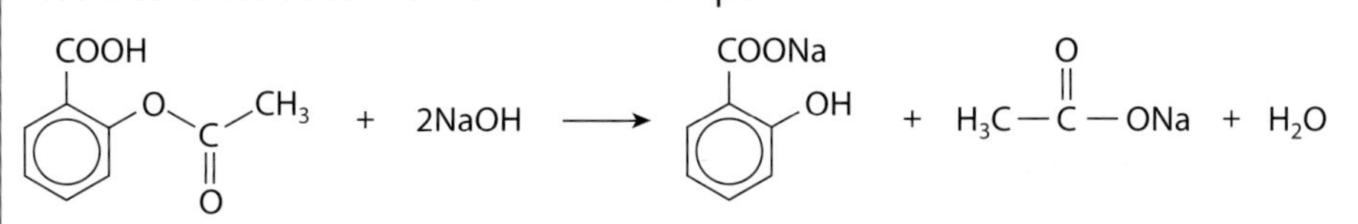

Since an excess of sodium hydroxide is used, the amount remaining is determined by titrating it against a standard solution of sulphuric acid. The reaction taking place is:

$2NaOH + H_2SO_4 \rightarrow 2H_2O + Na_2SO_4$

The difference between the initial and excess amounts of sodium hydroxide allows the mass of aspirin in the tablet to be calculated.

A standard solution is one of accurately known concentration.

UNIT 3 PPA 5 – ASPIRIN DETERMINATION contd

Procedure

Three aspirin tablets with an approximate mass of 1·5 g were added to a conical flask.

25·0 cm^3 of 1·00 mol l^{-1} sodium hydroxide solution was pipetted into the flask along with approximately 25 cm^3 of water. The flask was placed on a hot plate and the mixture was simmered gently for about 30 minutes. After cooling, the hydrolysed solution was transferred to a 250 cm^3 standard flask along with rinsings. The solution was made up to the graduation mark with water and, after stoppering the flask, it was inverted several times to ensure the contents were thoroughly mixed. 25·0 cm^3 of this solution was pipetted into a conical flask along with a few drops of phenolphthalein indicator and titrated against 0·0500 mol l^{-1} sulphuric acid. The end-point of the titration was indicated by the colour change, pink to colourless. The titrations were repeated until concordant results were obtained.

DON'T FORGET

To work out the mean titre volume, you average the concordant titre volumes. You must not include the rough titre volume.

DON'T FORGET

You only used 25 cm^3 portions of the hydrolysed aspirin solution in the titrations, so you must scale up your answer to find the number of moles of sodium hydroxide left in the original 250 cm^3 of hydrolysed solution.

Results

Titre		Trial	1	2
Burette readings/cm^3	Initial	1·5	17·1	0·6
	Final	17·1	32·3	15·7
Titre volume/cm^3		15·6	15·2	15·1
Mean titre volume/cm^3	15·15			

Number of moles of sulphuric acid used in titration = 0·01515 × 0·0500 = $7{\cdot}575 \times 10^{-4}$ mol

1 mole of sulphuric acid reacts with 2 moles of sodium hydroxide.

Therefore, number of moles of sodium hydroxide left in 25·0 cm^3 of the hydrolysed solution
= $2 \times 7{\cdot}575 \times 10^{-4} = 1{\cdot}515 \times 10^{-3}$ mol

Number of moles of sodium hydroxide left in 250·0 cm^3 of the hydrolysed solution
= $10 \times 1{\cdot}515 \times 10^{-3} = 1{\cdot}515 \times 10^{-2}$ mol

Number of moles of sodium hydroxide added to aspirin tablets initially
= $0{\cdot}0250 \times 1{\cdot}00 = 2{\cdot}50 \times 10^{-2}$ mol

Number of moles of sodium hydroxide which reacted with aspirin
= $2{\cdot}50 \times 10^{-2} - 1{\cdot}515 \times 10^{-2} = 9{\cdot}85 \times 10^{-3}$ mol

2 moles of sodium hydroxide reacts with 1 mole of aspirin.

Therefore, number of moles of aspirin in 3 tablets = $\frac{9{\cdot}85 \times 10^{-3}}{2} = 4{\cdot}925 \times 10^{-3}$ mol

Number of moles of aspirin in 1 tablet = $\frac{4{\cdot}925 \times 10^{-3}}{3} = 1{\cdot}642 \times 10^{-3}$ mol

Mass of aspirin in each tablet = n × GFM = $1{\cdot}642 \times 10^{-3} \times 180{\cdot}0$ = 0·296 g = 296 mg

Conclusion

The mass of aspirin per tablet was determined to be 296 mg.

Evaluation

The mass of aspirin in the tablet (296 mg) compares well with the manufacturer's specification (300 mg). To improve the reliability of the result, the experiment should be duplicated.

It is important that plain aspirin tablets are used in this determination because other types may contain ingredients which react with the sodium hydroxide solution added initially. If this were the case, then the mass of aspirin per tablet would be overestimated.

contd

UNIT 3 PPA 5 – ASPIRIN DETERMINATION contd

The calculation of the uncertainty in the mass of aspirin is detailed below.

Uncertainty in initial volume of NaOH (25 cm^3 pipette) $= 0{\cdot}06\,cm^3$

% uncertainty in initial volume of NaOH $= \frac{0{\cdot}06}{25{\cdot}0} \times 100 = 0{\cdot}24\%$

Uncertainty in concentration of NaOH (provided by teacher) $= 0{\cdot}008\,mol\,l^{-1}$

% uncertainty in concentration of NaOH $= \frac{0{\cdot}008}{1{\cdot}00} \times 100 = 0{\cdot}80\%$

% uncertainty in initial moles of NaOH $= 0{\cdot}24 + 0{\cdot}80 = 1{\cdot}04\%$

Hence, absolute uncertainty in initial moles of NaOH $= \frac{1{\cdot}04}{100} \times 0{\cdot}0250 = 2{\cdot}60 \times 10^{-4}\,mol$

Uncertainty in 250 cm^3 of hydrolysed aspirin solution (250 cm^3 standard flask) $= 0{\cdot}30\,cm^3$

% uncertainty in 250 cm^3 of hydrolysed aspirin solution $= \frac{0{\cdot}30}{250{\cdot}0} \times 100 = 0{\cdot}12\%$

Uncertainty in 25 cm^3 of hydrolysed aspirin solution (25 cm^3 pipette) $= 0{\cdot}06\,cm^3$

% uncertainty in 25 cm^3 of hydrolysed aspirin solution $= \frac{0{\cdot}06}{25{\cdot}0} \times 100 = 0{\cdot}12\%$

Uncertainty in titre volume of H_2SO_4 (50 cm^3 burette) $= 0{\cdot}05 + 0{\cdot}05 = 0{\cdot}10\,cm^3$

% uncertainty in titre volume of H_2SO_4 $= \frac{0{\cdot}10}{15{\cdot}15} \times 100 = 0{\cdot}66\%$

Uncertainty in concentration of H_2SO_4 (provided by teacher) $= 0{\cdot}0004\,mol\,l^{-1}$

% uncertainty in concentration of H_2SO_4 $= \frac{0{\cdot}0004}{0{\cdot}0500} \times 100 = 0{\cdot}80\%$

% uncertainty in moles of NaOH left $= 0{\cdot}12 + 0{\cdot}12 + 0{\cdot}66 + 0{\cdot}80 = 1{\cdot}70\%$

Hence, absolute uncertainty in moles of NaOH left $= \frac{1{\cdot}70}{100} \times 0{\cdot}01515 = 2{\cdot}58 \times 10^{-4}\,mol$

Absolute uncertainty in moles of NaOH reacting with aspirin
$= 2{\cdot}60 \times 10^{-4} + 2{\cdot}58 \times 10^{-4} = 5{\cdot}18 \times 10^{-4}\,mol$

% uncertainty in moles of NaOH reacting with aspirin $= \frac{5{\cdot}18 \times 10^{-4}}{9{\cdot}85 \times 10^{-3}} \times 100 = 5{\cdot}30\%$

This will be equal to the % uncertainty in the mass of aspirin.

Hence, absolute uncertainty in mass of aspirin per tablet $= \frac{5{\cdot}30}{100} \times 296 = 16\,mg$

So, mass of aspirin per tablet = 296 ± 16 mg

It is evident from this calculation that the major factors contributing to the uncertainty in the mass of aspirin per tablet are the percentage uncertainties in the concentrations of the sodium hydroxide (0·80%) and sulphuric acid (0·80%) and the percentage uncertainty in the titre volume of sulphuric acid (0·66%).

DON'T FORGET

When a calculation involves a multiplication or a division, you add the **percentage uncertainties** in the individual measurements.

DON'T FORGET

When a calculation involves an addition or subtraction you add the **absolute uncertainties** in the individual measurements.

THE INVESTIGATION REPORT

The Chemical Investigation is the equivalent of half a unit in Advanced Higher Chemistry. You have to pass a NAB which involves keeping a Day Book (see below) and you have to write an investigation report which should be in the region of 2500 words. The report is marked by an SQA marker out of a total of 25 marks. Your investigation mark is added to your mark out of 100 for the exam paper, giving you a total mark out of a maximum of 125 marks. This mark is then converted into your grade for Advanced Higher Chemistry.

SOURCES OF IDEAS FOR YOUR INVESTIGATION

Your best source of ideas is your chemistry teacher or tutor as he or she will know which investigations have been successful at your centre in previous sessions. Your teacher will also be aware of what apparatus and chemicals are available to you in your centre.

'School Science Review', published by the ASE, and 'Education in Chemistry', published by the Royal Society of Chemistry, are also very useful sources of ideas.

Your school will also have copies of 'Starter Investigations for Advanced Higher Chemistry' and 'More Starter Investigations for Advanced Higher Chemistry', written by K. Robertson and C. Gray and published in 2000 by the Higher Still Development Unit.

You may also find some other sources of ideas on the internet. The Heriot-Watt 'Scholar programme' on-line tutor may also be very helpful, providing ideas and advice as you carry out your Investigation.

GETTING STARTED

You have now decided on a topic for your chemical investigation. You must finalise the aim (or aims) of the investigation. Then you need to do some planning.

Find out:

- how your aim(s) can be achieved
- what apparatus and chemicals are needed
- if these materials are available
- any potential risks involved
- the timescale involved
- what information is available to you from books in your centre and if you can get information from local colleges, universities or product manufacturers
- information on the internet.

DON'T FORGET

Your teacher is likely to be the best source of information and the person you should seek advice from before, during and after you have carried out the practical work in your investigation.

KEEPING A DAY BOOK

Bear in mind that you may be asked to send your Day Book to the SQA if your centre is selected for verification of the Chemical Investigation NAB. Like any unit, there are certain Outcomes and Performance Criteria (PC) which are assessed. These are covered by keeping your Day Book up to date.

Outcome 1 is about developing a plan for your Investigation and it has three performance criteria. These are:

(a) You must record in a regular manner what you are doing in your investigation.
Your record should be in the form of a Day Book (a jotter or notebook, for example). You should write regular notes into your Day Book, especially when you are in the lab working on your investigation. These notes can be brief and should include ideas you have had, including any you rejected, and any contributions made by others, such as your teachers. You should write down any sources of information such as books, journals and websites. For websites you should also note down the date you accessed them as you will need this information later when you write your investigation report.

(b) You must state clearly the aims of your investigation.
The overall aims of your investigation and the aims of each experiment should be noted. You may change your original aims as you do the experiments during the investigation. You should record any changes and reasons for these changes in your Day Book.

(c) You need to plan which experimental techniques and apparatus, including chemicals, are appropriate to your investigation.
You should comment on how and why you chose a particular method or design for your practical work. This may include carrying out a risk assessment. The result of the risk assessment will tell you whether:

- the method and chemicals you have chosen are safe enough for you to proceed
- you need to wear safety goggles, gloves, aprons and do some of the practical work in a fume cupboard
- you have to reduce the quantities of some chemicals to make the practical work safer
- you have to substitute safer chemicals or methods in place of your original choices
- you have to abandon your original method or chemicals and look for something safer
- you have to take particular care when disposing of some of these chemicals.

A very useful website for safety information about chemicals which will help you complete your risk assessment is http://msds.chem.ox.ac.uk/#MSDS

DON'T FORGET

You need to carry out a risk assessment before you start the practical work.

Outcome 2 is about collecting and analysing the information you obtained in the practical work associated with your investigation. It also has three performance criteria. These are:

(a) You must carry out the experiments and obtain results accurately.
It is your investigation; you should be the person doing the experiments and obtaining results. Beware of going to a nearby college or university to do your investigation and getting much of the practical work done by a technician. You will not get marks for experimental work that you have not done yourself.

(b) You must record your measurements and observations in an appropriate format.
Your results, such as titration results, must include your raw data such as initial and final burette readings and not just your titre values. These should be recorded clearly in tables with correct headings and units. Make sure you record any observations, for example colours of solutions and colour changes, the formation of a precipitate, shapes and colours of any crystals formed and bubbles of gas being given off. Digital photographs can be used to record apparatus and also results such as chromatograms before and after you have used them to calculate R_f values.

contd

All these performance criteria will be covered by keeping your Day Book up to date with your plans, results and observations, and by getting your teacher to check it regularly.

KEEPING A DAY BOOK contd

(c) You must analyse your results and present these in an appropriate format.
Examples of appropriate formats include tables, graphs and diagrams. Calculations should be included. You may also wish to consider sources of error and estimates of error in measuring instruments. If your investigation included the preparation of a substance, you may have calculated the percentage yield; you may wish to note the factors which might have affected the yield, such as mass transfer losses, impurities in the reactants, equilibrium reactions and so on. You may have measured the melting point of a substance and could note information from reference data that helps to identify the substance. You may find it easier to do this in the evaluation part of your investigation report (print off a copy and put it into your Day Book for verification).

WRITING YOUR REPORT

Your report should be logical and easy to read. It should be written in such a way and with sufficient detail that someone who has read your report could repeat what you did. Every year the SQA produces guidelines on how to write up your investigation report. You may get a copy from your teacher but you can download it from the SQA website.

You must use the most up-to-date information on how to write the investigation report.

Find guidance on writing your investigation report at http://www.sqa.org.uk/sqa/controller?p_service=Content.show&p_applic=CCC&pContentID=39860&search Qtext=Advanced+Higher+%3E+Chemistry and then click onto 'Investigation guidance'.
Following the information given in this guidance document will almost certainly give you extra marks. On the other hand, if you don't use the guidance document, you will definitely lose marks.

HOW WILL THE INVESTIGATION REPORT BE MARKED?

Many candidates lose marks because they have not followed the information on how to cite and list references exactly as shown in the guidance document.

Category 1 – Presentation (3 marks)

- The first mark is for title page, contents page, page numbers and **at least** three references correctly cited in the text and correctly listed at the back of the report. These references can be books, journals and websites but must be recorded **exactly** as shown in the SQA guidance document. Note that the method of citing a source in the main body of the text is totally different from how the reference is listed at the back of the report. This may also be different from how you are expected to write references in your physics or biology reports.
- You must put in a brief summary or **abstract** immediately after the contents page and under a separate heading, preferably on a separate page. This summary must state the **main aim(s) and overall finding(s)** of your investigation. Each aim stated here must be covered in the **conclusions** at the end.
- Your report must be clear and concise; it should be easy to read and understand. It may be easier for the marker to follow your report if the experimental results are given after each procedure. It also helps if you put in an aim for each separate experiment so that the marker knows what you are doing and why.

contd

HOW WILL THE INVESTIGATION REPORT BE MARKED? contd

Category 2 – Underlying Chemistry (4 marks)

There are four marks for the **underlying chemistry**; it is worth spending time to get the maximum possible number of marks. The marks are awarded for appropriate use of formulae, equations, structures and background theory, for example. You could put in some interesting historical information here, but you will get no marks for it.

- Think about the chemistry of your experiments and include details in your report. This is particularly important if your investigation covers work done in the Higher or Advanced Higher course that you should know well and be able to explain.
- Try to cover the underlying chemistry in your own words.
- Do not copy from a textbook or 'cut and paste' from a website.
- Write your investigation report in your own words; use words that are easily understood rather than words you do not know.
- Explain the meaning of any new or unfamiliar words.

DON'T FORGET

There are four marks for underlying **chemistry** and the marker will be looking purely for chemical information.

Category 3 – Procedures (6 marks)

The procedures must be written in the **past tense** and **impersonal passive voice**. The six marks are awarded in the following way:

- One mark for the procedures being appropriate to the aim(s) of your investigation.
- One mark for sufficient detail to allow the investigation to be repeated by another Advanced Higher Chemistry student using only your report as guidance. You will lose a mark here if you use the words 'I', 'we', 'my', or if you use the present tense. Likewise, you will lose the mark if you give the procedure for any experiment as a set of instructions.
- One mark for the investigation being at the correct level for Advanced Higher Chemistry.
- One mark for creativity or originality on your part, such as using more than one technique **or** for making a modification to your technique in the light of experience.
- One mark for duplicating your experiments. You should mention that you have duplicated each experiment and make sure the results show that the experiments have been duplicated.
- One mark for the accuracy of your measurements or for using the correct apparatus. For example, you would lose this mark if you were using a measuring cylinder to dispense a liquid when you should be using a pipette or burette for accuracy. So, make sure the marker knows what pieces of equipment you used for each experiment.

Note that repeating titrations to get concordant results is not duplication.

You need to show the marker that you have carried out each experiment in duplicate to improve the reliability of your results.

Category 4 – Results (5 marks)

The subdivision of these marks depends on whether the marker feels that your investigation has been quantitative or qualitative. The marker will mark this category in such a way that you benefit from his or her decision.

Quantitative investigations

- One mark for your results being relevant to the aim(s) and for the raw data being properly recorded. Raw results should be recorded for each experiment. If you use a balance and tare it, you should make it clear that this is how you measured the mass of any substance.
- One mark for your results being within the correct limits of accuracy of measurement. For example, all masses recorded to 2 or 3 decimal places and burette readings to at least 1 decimal place. Be wary of having too few or too many significant figures in your raw and processed results, and in your final results.

When recording titration results include initial and final readings. Don't just put in processed titre results.

contd

HOW WILL THE INVESTIGATION REPORT BE MARKED? contd

Remember a graph should normally be a line of best fit. Don't just join the points plotted.

- Two marks for your results (raw and processed) being presented in a clear and concise manner. The marker will be looking for tables, graphs, diagrams and calculations. Graphs and diagrams must be labelled correctly, have correct headings, with lines drawn accurately and appropriate scales. You do not have to show each and every calculation, but you must give an example of **each type of calculation** with a set of your results.
- One mark for descriptions of observations you have made, such as colour changes, colour of precipitates, bubbles or shapes of crystals. Make sure you put in lots of observations.

Qualitative investigations

- Two marks for recording your results to the correct limits of accuracy, as above. Your results should be presented clearly using an appropriate format. Raw data may include chromatograms, photographs or diagrams of results. Your results may also include masses of products, melting point measurements, yield and percentage yield of products.
- Three marks for descriptions of your observations. These would include colours of solutions, colour changes, shapes and colours of crystals formed, colours of any precipitates, results of tests and so on.

Make sure you cover all your aims in your overall conclusions, otherwise you lose a mark.

Category 5 – Conclusions and Evaluation (7 marks)

This is the most discriminating category.

- The first two marks are for your conclusion(s). The first mark is for all the aims being covered in your conclusion; the conclusion(s) must relate to all the aim(s) given in the summary or abstract at the start of your report. The second mark is for your conclusions being valid in relation to your results and your calculations.

DON'T FORGET

This part is worth four marks. You should consider any errors in your procedures, how these might have been reduced and what effect they had on your final results.

- There are four marks awarded for the evaluation of your procedures and evaluation of your results. This is an opportunity for you to take an overview of what you have done in your investigation and to consider the good and bad points. You should consider how your investigation might have been improved, together with the main sources of error in your procedures and how these affected your results. For example, consider the accuracy of the equipment you used and how this affected the results. Discuss how impurities in reactants would have made a difference. Highlight any modifications you made during the investigation and say which modifications you would make if you were to start again. You should mention whether you carried out experiments in duplicate and the reasons for doing this. You may choose to carry out uncertainty calculations. See examples on pages 53 and 93.
- The final mark is a quality (or bonus) mark based on the **overall quality of your investigation.** To be eligible for this you must have scored at least three marks for Underlying Chemistry and at least three marks for your Evaluation.

ANSWERS

p5

Energy = 235 kJ mol^{-1}

p7

Energy = 1313 kJ mol^{-1} which compares well with Data Booklet value of 1311 kJ mol^{-1}

p9

There will be 14 electrons which can go into the f subshell and so there will be seven f orbitals.

p11

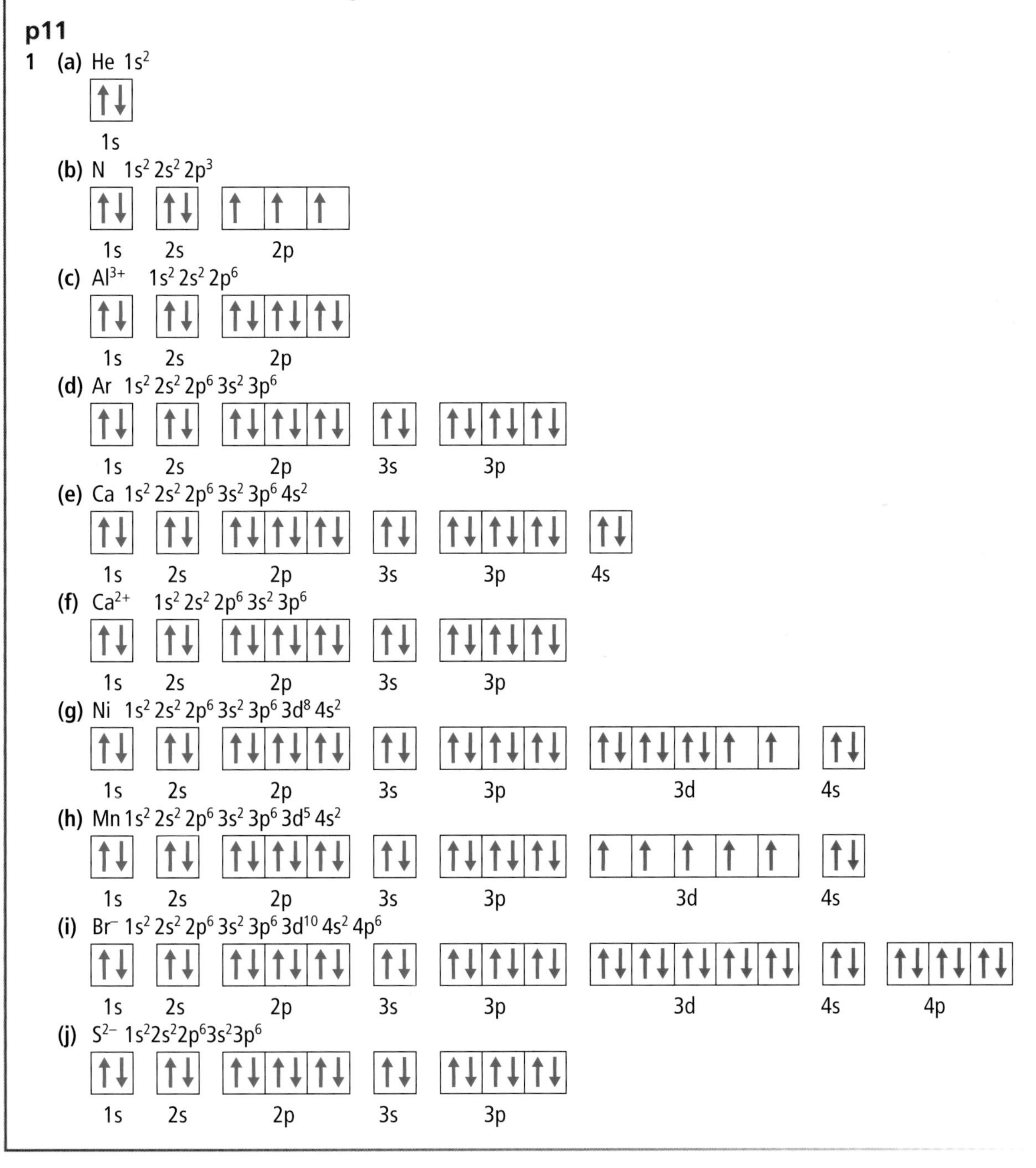

2 The electronic configuration of beryllium is $1s^2 2s^2$ and for boron it is $1s^2 2s^2 2p^1$.
The electron to be removed from a beryllium atom is from a full 2s **subshell** and because full subshells are fairly stable, more energy is required to remove this electron than is required to remove the outer electron from a boron atom. Therefore beryllium has a slightly higher first ionisation energy.
The electronic configuration of nitrogen is $1s^2 2s^2 2p^3$ and for oxygen it is $1s^2 2s^2 2p^4$.
The electron to be removed from a nitrogen atom is from a half-full 2p **subshell** and because half-full **subshells** are fairly stable, more energy is required to remove this electron than is required to remove the outer electron from an oxygen atom. Therefore nitrogen has a slightly higher first ionisation energy.

p13

1 By using a spectroscope to look indirectly at the light from the sun. Lines which did not correspond to any known element were present in the spectrum and so a new element was discovered and it was named after the Greek word for the sun, "Helios".

2 $1{\cdot}0 \times 10^{-4}$ mol l^{-1}

p15

1 As the hydrogen atoms move closer each positive nucleus attracts the negative electron of the other atom and the potential energy drops. Point D shows the optimum distance between the two nuclei taking into account the electrostatic attractions between the positive nuclei and the negative electrons of the other atom. This distance r_0 can be considered the H – H bond length. Between D and E the two nuclei are becoming too close and the large increase in potential energy is due to the repulsions between these positive nuclei.

2 Molten sodium hydride will conduct electricity and electrolysis would produce sodium at the negative electrode and hydrogen at the positive electrode.

p17

1 (a)

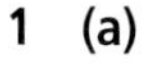

(b)

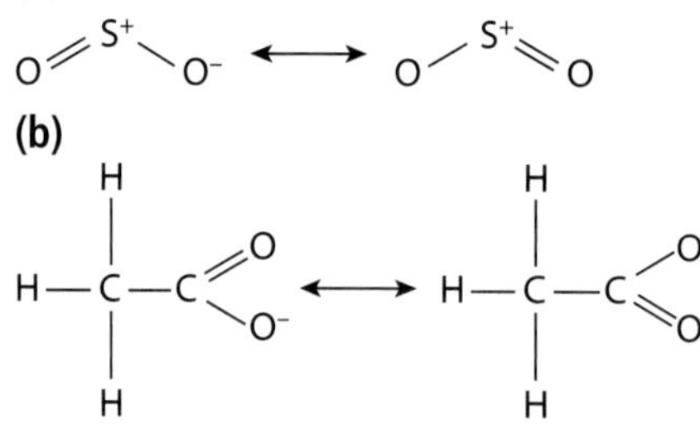

2 **(a)** 4 bonding pairs, 0 non-bonding pairs so shape is tetrahedral (like methane)
(b) 3 bonding pairs, 1 non-bonding pair so shape is pyramidal (like ammonia)
(c) 2 bonding pairs, 2 non-bonding pairs so shape is non-linear or bent (like water)
(d) 3 bonding pairs and 2 non-bonding pairs so molecule is shaped like the letter "T"
(e) 2 bonding pairs and 3 non-bonding pairs so shape is probably linear

p19

1

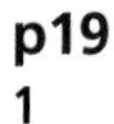

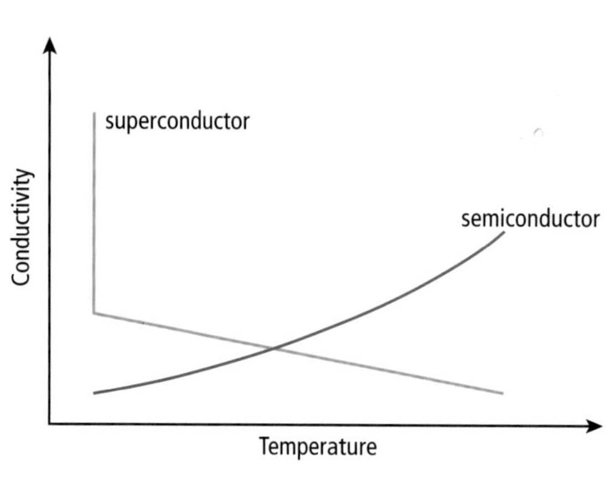

2 **(a)** Li^+F^- = 68/133 = 0·51 so NaCl structure
(b) $Mg^{2+}O^{2-}$ = 65/136 = 0·48 so NaCl structure
(c) $Ba^{2+}O^{2-}$ = 135/136 = 0·99 so CsCl structure.

p21

1 D 2 D 3 D 4 C 5 D

p23

1 Scandium always forms Sc^{3+} ions and since the 4s electrons are lost first, then the electronic configuration of the Sc^{3+} ion is $1s^22s^22p^63s^23p^6$ so no electrons in the d subshell.
Zinc always forms Zn^{2+} ions and the electronic configuration of the Zn^{2+} ion is $1s^22s^22p^63s^23p^63d^{10}$ and it does not have an **incomplete** subshell.
2 The electronic configuration for Fe^{2+} is $1s^22s^22p^63s^23p^63d^6$ and for Fe^{3+} it is $1s^22s^22p^63s^23p^63d^5$ in spectroscopic notation.
In orbital box notation, considering only the d orbitals

↑↓	↑	↑	↑	↑

Fe^{2+}

↑	↑	↑	↑	↑

Fe^{3+}

There is a special stability associated with all the orbitals in the d subshell being half-filled and so Fe^{3+} is more stable than Fe^{2+}.
3 Oxidation number is 6 or VI and since Cr is in a high oxidation state then $Cr_2O_7^{2-}$ is a good oxidising agent. Usually when it acts as an oxidising agent it is reduced to the Cr^{3+} ion.

p25

1 **(a)** tetrachlorocobaltate(II)
(b) hexaamminenickel(II)
(c) hexachloroplatinate(IV)

p27

1 Chromium is in oxidation state (VI) in $Cr_2O_7^{2-}$. The electronic configuration of chromium atoms is $1s^2\,2s^2\,2p^6\,3s^2\,3p^6\,3d^5\,4s^1$ and so the electronic configuration of Cr(VI) is $1s^2\,2s^2\,2p^6\,3s^2\,3p^6$ which means that there are no electrons in the d subshell and so d – d transitions are not possible.
2 **(i)** The $[CoCl_4]^{2-}$ complex is absorbing light mainly from the red and also partly from the orange region of the visible spectrum and this explains the blue-green or cyan colour of this complex in solution. The $[Co(H_2O)_6]^{2+}$ complex is pink in solution because it is absorbing light mainly from the middle of the visible spectrum
(ii) The energy differences between the low energy d orbitals and the higher energy d orbitals will be different in the two complexes because of the different positions of the Cl^- and H_2O ligands in the spectrochemical series. Also the $[CoCl_4]^{2-}$ complex is tetrahedral and the $[Co(H_2O)_6]^{2+}$ is octahedral in shape and this too causes differences in the splitting of the d orbitals.

p31

1 Mass of water driven off = 2·58 – 2·22 = 0·36 g
Mass of $BaCl_2$ = 2·22 g

Number of moles of H_2O = 0·36/18 = 0·020 mol
Number of moles of $BaCl_2$ = 2·22/208.3 = 0·011 mol

Ratio of moles $BaCl_2 : H_2O$ = 0·011:0·020 = 1:1·8 which is 1:2 to the nearest whole number.
Therefore n = 2 and the formula of hydrated barium chloride is $BaCl_2.2H_2O$.

2 You are given the volume and concentration of the sulphuric acid, so the number of moles of sulphuric acid, $n = V \times c = 0{\cdot}0178 \times 0{\cdot}22 = 0{\cdot}003916$ mol.
The balanced stoichiometric equation for the reaction shows us that 2 mol of NaOH reacts with 1 mol of H_2SO_4:

$2NaOH$	+	H_2SO_4	$\rightarrow$	Na_2SO_4	+	$2H_2O$
2 mol		1 mol				
$2 \times 0{\cdot}003916$ mol		$0{\cdot}003916$ mol				

Therefore in the $25{\cdot}0\,cm^3$ sample of the diluted drain cleaner there was $2 \times 0{\cdot}003916 = 0{\cdot}007832$ mol of NaOH.
In the $250\,cm^3$ standard flask, there must have been $0{\cdot}007832 \times 10 = 0{\cdot}07832$ mol of NaOH
So in the $10{\cdot}0\,cm^3$ of the undiluted drain cleaner there was $0{\cdot}07832$ mol of NaOH
In one litre ($1000\,cm^3$) there would be $0{\cdot}07832 \times \frac{1000}{10} = 7{\cdot}832$ mol of NaOH
The mass of NaOH $= n \times FM = 7{\cdot}832 \times 40{\cdot}0 = 313$ g in one litre.

p33

1 The equilibrium equation is $H_2O(l) \rightleftharpoons H^+(aq) + OH^-(aq)$

(a) $K = [H^+(aq)][OH^-(aq)]$ (Remember that $[H_2O]$ is usually omitted from the expression for the equilibrium constant)

(b) Since K is increasing, then the position of equilibrium must be shifting to the right as the temperature increases. Increasing the temperature favours the endothermic reaction and so the forward reaction must be endothermic and so $\Delta H^{\ominus}$ must be +ve.

p35

In the extraction using $100\,cm^3$ of CH_2Cl_2 in one extraction:
Let x be the mass of caffeine extracted from the aqueous tea solution into the CH_2Cl_2.
Therefore the $[\text{caffeine(aq)}] = \frac{0{\cdot}30 - x}{100}$ and
$[\text{caffeine}(CH_2Cl_2)] = \frac{x}{100}$
Substituting these values into the expression,
$K = \frac{[\text{caffeine}(CH_2Cl_2)]}{[\text{caffeine(aq)}]} = 4{\cdot}6$ and solving for x gives a
result of **<u>0·246</u> g** of caffeine extracted using $100\,cm^3$ of CH_2Cl_2 in one extraction.
In the two consecutive extractions $50\,cm^3$ of CH_2Cl_2 each time:
Let y be the mass of caffeine extracted from the aqueous tea solution into the first $50\,cm^3$ of CH_2Cl_2.
Therefore the $[\text{caffeine(aq)}] = \frac{0{\cdot}30 - y}{100}$ and
$[\text{caffeine}(CH_2Cl_2)] = \frac{y}{50}$
Substituting these values into the expression,
$K = \frac{[\text{caffeine}(CH_2Cl_2)]}{[\text{caffeine(aq)}]} = 4{\cdot}6$ and solving for y gives a
result of **<u>0·209</u> g** of caffeine extracted using $50\,cm^3$ of CH_2Cl_2 in the first extraction.
The mass of caffeine remaining in the aqueous solution of tea is now 0·091 g
(from 0·30 – 0·209)
Let z be the mass of caffeine extracted from the aqueous tea solution into the second $50\,cm^3$ of CH_2Cl_2.
Therefore the $[\text{caffeine(aq)}] = \frac{0{\cdot}091 - z}{100}$ and
$[\text{caffeine}(CH_2Cl_2)] = \frac{z}{50}$
Substituting these values into the expression,
$K = \frac{[\text{caffeine}(CH_2Cl_2)]}{[\text{caffeine(aq)}]} = 4{\cdot}6$ and solving for z gives a
result of **<u>0·063</u> g** of caffeine extracted using $50\,cm^3$ of CH_2Cl_2 in the second extraction.

So the total mass of caffeine extracted using two 50 cm^3 samples of CH_2Cl_2 = 0·209 + 0·063 = **0·272 g** of caffeine which is greater than the 0·246 g of caffeine obtained using the single extraction method.

So using the two 50 cm^3 samples of CH_2Cl_2, **0·026 g** more caffeine is extracted.

p37

1 $H_3O^+(aq)$ is the conjugate acid and $CH_3COO^-(aq)$ is the conjugate base.

2 (a) $[H^+] = 0{\cdot}22$ mol l^{-1} and so pH = –log(0·22) = 0·66

(b) $[OH^-] = 0{\cdot}12$ mol l^{-1} and so $[H^+] = \frac{10^{14}}{0{\cdot}12} = 8{\cdot}33 \times 10^{-14}$ mol l^{-1}

so pH = $-\log 8{\cdot}33 \times 10^{-14} = 13{\cdot}1$

3 (a) pH = 4·3, so $-\log[H^+] = 4{\cdot}3$ and $\log[H^+] = -4{\cdot}3$, therefore $[H^+] = 5{\cdot}0 \times 10^{-5}$ mol l^{-1}

(b) $[H^+] = 5{\cdot}0 \times 10^{-9}$ mol l^{-1}

4 (a) $pH = \frac{1}{2}pK_a - \frac{1}{2}\log c = \frac{1}{2}(4{\cdot}76) - \frac{1}{2}\log 0{\cdot}1 = 2{\cdot}88$

(b) pH = 2·45

p39

1 (a) Phenolphthalein, since its pH range of 8·0 – 10·0 fits into the "vertical" part of the weak acid/strong alkali graph

(b) Any one from methyl red, bromocresol green or methyl orange since their pH ranges fit into the vertical part of the strong acid/weak alkali graph.

2 A salt made from propanoic acid and a strong base would be suitable so acceptable answers include sodium propanoate and potassium propanoate.

3 The relevant equations are

$HCOOH(aq) \rightleftharpoons HCOO^-(aq) + H^+(aq)$ and

$Na^+HCOO^-(s) \rightarrow Na^+(aq) + HCOO^-(aq)$

(a) When an acid is added, the H^+ ions from the acid react with the $HCOO^-$ ions **from the sodium methanoate salt** to form HCOOH molecules so the $[H^+]$ remains the same as before and the pH remains constant.

(b) When an alkali is added, the OH^- ions from the alkali react with the H^+ ions from the dissociation of the methanoic acid. To compensate for this, more methanoic acid molecules dissociate to replace these H^+ ions so, once again, the $[H^+]$ remains the same as before and the pH remains constant.

(c) When water is added the ratio $\frac{[acid]}{[salt]}$

is not altered and so the pH value remains constant.

4 Using the expression, $pH = pK_a - \log \frac{[acid]}{[salt]}$ the pH is calculated as 4·94.

(Remember that 40 cm^3 of 0·1 mol l^{-1} ethanoic acid diluted to 100 cm^3 becomes 0·04 mol l^{-1} and 60 cm^3 of 0·1 mol l^{-1} sodium ethanoate diluted to 100 cm^3 becomes 0·06 mol l^{-1})

p41

1 (a) –279 kJ mol^{-1} (b) 226 kJ mol^{-1}

2 (a) –257.5 kJ mol^{-1} (b) 199 kJ mol^{-1}

3 The answers to **2** were calculated using **mean** bond enthalpy values but the bond enthalpy values of the actual bonds in ethanol and ethyne may well be different from the mean values quoted in the SQA Data Booklet.

p43

1 This is because energy is being released when the negative electron is attracted by the positive nucleus into the outer shell and a stable electron arrangement results. The electron affinity is a measure of the attraction between the incoming electron and the nucleus - the stronger the attraction, the more energy is released. The factors which affect this attraction are exactly the same as those relating to ionisation energies - nuclear charge, distance between the positive nucleus and the incoming electron and screening or shielding.

2 In this case the incoming electron is being "forced" into an negative ion. The positive value indicates that energy is required to do this. The second electron affinity of oxygen is high because the electron is being forced into a small, very electron-dense space.

3 $-385{\cdot}5\,kJ\,mol^{-1}$

4 Answer is $-155\,kJ\,mol^{-1}$. See the thermochemical cycle below.

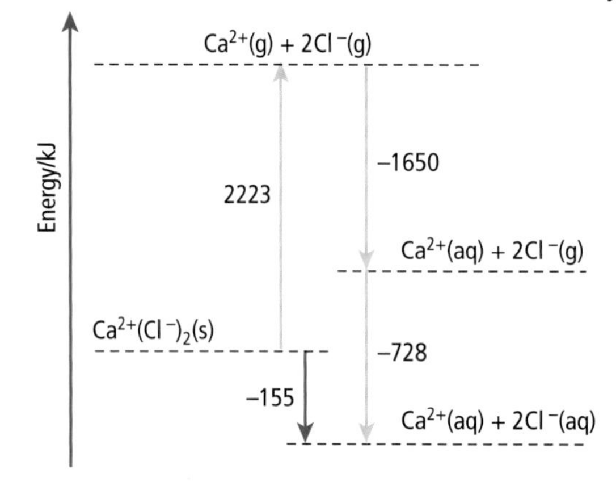

p47

Equilibrium is reached when approximately 75% of R is converted into P.

Hence $K = \frac{[\text{products}]}{[\text{reactants}]} = \frac{75}{25} = 3$

p49

$\Delta G^\circ = -nFE^\circ = -4 \times 9{\cdot}65 \times 10^4 \times 1{\cdot}23 = -4{\cdot}75 \times 10^5\,J\,mol^{-1} = -475\,kJ\,mol^{-1}$

p51

x = 2 and **y** = 0.50

p65

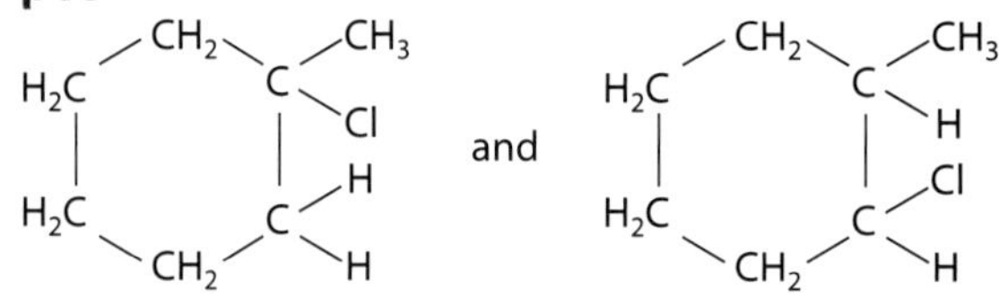

The major product is called 1-chloro-1-methylcyclohexane.

p75

D since $C_2H_5N(CH_3)_2$ is a tertiary amine.

p77

(a) concentrated nitric and sulphuric acids

(b) ethanoic acid or ethanoyl chloride or ethanoic anhydride

(c) **(i)** electrophilic substitution

(ii) reduction

(iii) condensation

(d) Phenylamine is a weaker base than methylamine because the lone pair of electrons on the nitrogen atom in phenylamine accepts a proton less readily than the lone pair in methylamine. The reason for this is that the lone pair of electrons on the nitrogen atom in phenylamine becomes part of the pi molecular orbital on the benzene ring making the phenylamine molecule more stable.

p85

(a) since the sulphanilamide molecule has roughly the same shape as 4-aminobenzoic acid.

(b) Sulphanilamide plays the role of antagonist since it binds to the receptor site and prevents 4-aminobenzoic acid from binding.

INDEX